TRUCK COMPANY FIREGROUND OPERATIONS
Second Edition

Harold Richman

Fire Chief (Retired)

*Past President
International Society of
Fire Service Instructors*

National Fire Protection Association
Batterymarch Park
Quincy, Massachusetts 02269

J2-2129992

Project Manager: Gene A. Moulton
Editor: Marion Cole
Production Designer: G. S. Stevens
Production Coordinator: Muriel E. McArdle

Copyright © 1986 National Fire Protection Association
All rights reserved
NFPA Catalog No. FSP-76A
ISBN 0-87765-317-8
Library of Congress Card No. 85-63839
Printed in the United States of America.

DEDICATION

"'Duty' is the noblest word in the English language."
— Robert E. Lee

During my 35 years with the fire service, I have seen on many occasions the strong devotion to duty by many fire fighters. Duties often carried out under extreme physical and mental stress resulted in saving lives or making a "good stop."

This book is dedicated to those fire fighters and to the "truckies," "house wreckers," "opera stars" and "shepherds" in all fire departments, especially to my former colleagues and present friends in the truck companies of Washington, D.C.; Pearl Harbor, Hawaii; Memphis, Tennessee; Silver Spring, Maryland; Fairfax County, Virginia; Prince George's County, Maryland; Chicago; New York City; and the Town of Menasha, Wisconsin; and to those members of other departments who over the last 3½ decades extended their invitations to me and allowed me to observe and to serve with truck crews in action.

CONTENTS

Dedication............................... v
Preface................................. ix
Acknowledgments........................ xi

1 INTRODUCTION........................ 1
Fire Spread............................ 2
   Convection........................... 2
   Radiation............................ 4
   Conduction........................... 5
   Flashover............................ 6
   Smoldering Fire and Backdraft......... 6
Truck Company Operations............... 7
Summary............................... 8

2 INITIAL ASSIGNMENTS................. 9
Tools and Personnel.................... 9
   Hand Tools.......................... 10
   Power Tools......................... 10
   Training............................ 10
   Tool Assignments.................... 11
Coverage............................. 12
   Front and Rear Coverage............. 12
   Truck Out of Quarters............... 14
   Other Aspects....................... 14
   Size-up Information................. 14
Apparatus Positioning................. 16
   Approach............................ 16
   Positioning......................... 17
Summary.............................. 18

3 RESCUE............................. 19
Chronology of Rescue Operations....... 20
   Before the Alarm.................... 20
   At the Alarm........................ 20
   On the Fireground................... 21
Rescue Considerations................. 22
   Residential Occupancies............. 23
   Industrial Occupancies.............. 25
   Hospitals, Schools and Institutions.. 25
   Retail Stores....................... 27
Search............................... 28
   Search Duties....................... 29
   Standard Search Procedure........... 29
   Search Techniques................... 34
Summary.............................. 36

4 VENTILATION TECHNIQUES............. 39
Basic Principles..................... 40
Natural Openings..................... 41
   Windows............................. 41
Natural Roof Openings................ 43
   Skylights........................... 44
   Roof Scuttles (Hatches)............. 47
   Ventilators......................... 48
   Machinery Covers.................... 52
   Elevator Houses..................... 52
   Air Shafts.......................... 53
   Prefire Inspection.................. 54
Cutting through Roofs................ 55
Forced Ventilation................... 56

Smoke Ejectors.................................. 56
Fog Streams..................................... 57
Summary.. 58

5 VENTILATION OPERATIONS............ 59
One- and Two-story Dwellings............. 59
One-story Dwellings............................. 59
Two-story Dwellings............................. 60
Attic Fires....................................... 61
Basement Fires.................................. 61
Multiple-use Residential and Office Buildings 61
Roof Operations................................. 62
Venting.. 64
Ground-floor Stores............................. 65
Adjoining Buildings............................. 66
Shopping Centers, Row Stores, and Other One-story Buildings................ 66
Roof Operations................................. 66
Attached Occupancies........................... 68
Ground-level Ventilation........................ 68
Basement Fires in Large Structures........ 68
Basement Venting................................ 68
First-floor Venting............................. 71
Roof Venting.................................... 71
Fire Resistant Structures................... 71
Window Venting.................................. 71
Stairway Venting................................ 72
Other Truck Duties.............................. 72
Using Elevators to Approach the Fire Floor.. 72
Smoldering Fires............................. 74
Indications..................................... 74
Backdraft....................................... 75
Venting... 76
Summary.. 77

6 CHECKING FIRE EXTENSION.......... 79
Interior Fire Extension..................... 80
Fire in Concealed Spaces........................ 80
Vertical Fire Spread............................ 81
Horizontal Fire Spread...................... 85
Open Interior Spread............................ 89
Exterior Exposures.......................... 90
Summary.. 91

7 FORCIBLE ENTRY........................... 93
Prefire Inspection.......................... 93
Size-up....................................... 95
The Fire Building............................... 95
Exposed Buildings............................... 96
Tools... 97
Cutting Tools................................... 98
Prying and Forcing Tools........................ 98
Lock Pullers.................................... 99
Explosive Tools................................. 99

Forcible Entry through Windows.......... 99
Double-hung Windows............................ 100
Casement Windows............................... 101
Other Windows.................................. 101
Forcible Entry through Doorways......... 102
Commercial Occupancies: Front................. 102
Commercial Occupancies: Rear.................. 104
Dwellings and Apartments...................... 106
Office Buildings............................... 108
Other Occupancies.............................. 109
Sidewalk Basement Entrances................... 109
Summary....................................... 110

8 AERIAL OPERATIONS..................... 113
Safe Procedure............................. 114
Rescue....................................... 115
Spotting the Turntable......................... 116
Raising the Aerial Unit........................ 118
Placing the Ladder or Platform................. 119
Imperfectly Spotted Turntable.................. 120
Removing Trapped People........................ 121
Removing Victims by Litter..................... 122
Lifeline Anchor................................ 124
Ventilation................................. 125
Removing Windows............................... 126
Venting with Streams........................... 126
Venting with an Aerial Ladder.................. 128
Hose Operations............................. 128
Lifting Personnel and Equipment................ 129
Using Hose as a Portable Standpipe............. 130
Summary....................................... 130

9 GROUND LADDERS.......................... 131
Ground Ladder Operations................. 131
Ladder Work.................................... 132
Handling Ground Ladders........................ 132
Safety... 135
Rescue....................................... 135
The Normal Raise............................... 135
The Emergency or Hotel Raise................... 137
Ground Ladders as Exits........................ 137
Bridging....................................... 138
Ventilation................................. 140
Using Ladders as Venting Tools................. 140
Advancing Hoselines........................ 141
Placing Ladders................................ 141
Raising Ladders from Roofs..................... 142
Climbing Ladders............................... 143
Positioning Fire Fighting Streams........ 143
Other Uses.................................. 144
Transporting the Injured....................... 144
Covering Weakened Areas........................ 144
Summary....................................... 145

10 SALVAGE ... 147
Protecting Building Contents ... 149
Salvage Covers ... 149
Covering Building Contents ... 150
Controlling the Water Flow ... 151
Catchalls ... 152
Removing Water from Buildings ... 154
Chutes ... 154
Drains ... 156
Toilets ... 156
Sewer Pipes ... 156
Openings in Walls ... 157
Pumps ... 158
Summary ... 159

11 ELEVATED STREAMS ... 161
Setting Up the Aerial Pipe ... 161
Spotting the Turntable ... 161
Developing the Water Supply ... 164
Rigging the Aerial Pipe ... 165
Aerial Streams for Fire Attack ... 167
Nozzles ... 167
Stream Placement ... 167
Wind and Thermal-Updraft Effects ... 168
Weakened Structures ... 169
Shutdown ... 170
Improper Use of Streams ... 170
Aerial Streams for Exposure Coverage ... 171
Outside Exposures ... 172
Exposure Hazards ... 172
Exposure Protection ... 173
Elevated Handlines ... 174
Operating from a Platform ... 174
Operating from an Aerial Ladder ... 175
Summary ... 175

12 CONTROL OF UTILITIES ... 177
Prefire Planning for Utility Control ... 177
Operating Controls ... 178
Building Codes ... 178
Forced-Air Systems ... 178
Forced-Air Blowers ... 178
Dampers ... 179
Heating Systems ... 179
Cooling Systems ... 180
Air Circulation Systems ... 180
Heating Units and Fuels ... 181
Oil Burners ... 181
Kerosene Heaters and Stoves ... 182
Gas Units ... 182
Electric Service ... 184
Main Power Switches ... 185
Elevators ... 186
Water Pipes ... 186
Water Shut-off Valves ... 186
Boilers and Heating Units ... 186
Summary ... 187

13 OVERHAUL ... 189
Preinspection ... 190
Personnel ... 190
Control of Personnel Movements ... 191
Work Assignments ... 192
Procedure ... 192
Indications of Rekindling ... 194
Areas of Possible Rekindling ... 195
Chemicals and Other Hazards ... 198
Searching for the Cause of Fire ... 200
Restoration and Protection ... 200
Summary ... 201
Glossary ... 203
Index ... 209

PREFACE

Throughout the country, the continuing construction of taller buildings, sprawling shopping centers, apartment complexes and industrial parks, and the deterioration of old buildings and neighborhoods complicate the fire fighting problem. In some areas, this has required establishment of truck companies for the first time; in other communities, the need for additional truck companies has been recognized.

Truck companies are used extensively where methods of combined engine and truck company operations are understood and expertly applied. However, proficiency in these operations is not easily attained. It is the purpose of this book to present sound truck company operational procedures for working structural fires.

This book is based on the author's involvement in fireground activities as a fire fighter, company officer, training officer and chief officer in six different fire departments. These departments covered urban, suburban and rural fire fighting operations. The author's participation in and observation of truck company operations in those departments, and information and advice obtained from officers and members of several other departments, have contributed to the contents of this book.

This is a companion book to *Engine Company Fireground Operations*. The two books provide overall basic fireground procedures required for effective fire fighting activities. The text is also compatible with NFPA audiovisual materials, standards, and other publications for the fire service.

The objectives of *Truck Company Fireground Operations* are to assist the majority of fire departments in developing and conducting effective operations. It does not teach manipulative skills nor cover problems and procedures necessary in those few departments with unique situations.

ACKNOWLEDGMENTS

It is a pleasure to thank those who have assisted in the development of this book.

For their encouragement and guidance in the content of this book I wish to thank Chief John R. Leahy, Bureau of Fire, Pittsburgh, Pennsylvania; Deputy Chief William C. Quirk (Retired), Fire Department, Minneapolis, Minnesota; Assistant Chief Greg Cleveland, Fire Department, Town of Menasha, Wisconsin; Mr. Dominic Catera, Fire Department, New York City; and Mr. George D. Post, a veteran of the Fire Department, New York City.

For their patience, hospitality and some excellent meals during the writing of this book I must thank the members of Truck Co. No. 22, Department of Fire Protection, Prince George's County, Maryland, and members of truck companies of Fairfax County, Virginia; Chicago; New York City; The Town of Menasha, Wisconsin; San Francisco; Los Angeles; and those of many other cities who so graciously gave me their time. I also thank Mr. Gene Moulton and Miss Marion Cole of the NFPA Editorial Department, as well as other NFPA experts whose suggestions have enhanced the value of these pages.

INTRODUCTION 1

A number of fire service officials have stated that the overall efficiency of a fireground operation is determined by the performance of the truck companies — often also called ladder companies. Others, in an analogy with the military, consider engine companies as the fire fighting infantry and truck companies as the engineers. These statements characterize the importance of truck company operations and the role of the truck company on the fireground.

Truck companies provide access to, and exits from, all parts of a fire building. Truck crews also are responsible for removing heat, smoke and gases to allow greater visibility and permit engine company personnel to move rapidly and safely within a fire building or exposed buildings. These examples do not by any means include all the duties of a truck company, but they do illustrate two important points about truck work.

First, truck work is required at every fire, regardless of who does it. In a fire department that includes one or more truck companies, the truck work is clearly the responsibility of those companies. When a department does not include truck companies, arrangements are often made to have neighboring truck companies respond to first alarms. But even if truck companies are not available within a department or neighboring departments, the truck work must be performed. In such situations, it is important that truck work be assigned to particular personnel; that these fire fighters be thoroughly familiar with the tools, skills and operations of truck work; and that they be trained and assigned as a team.

Second, truck operations either accompany or precede engine operations. In many fire situations, attack lines cannot be advanced until some truck work has been done. If truck operations are performed inefficiently, other fireground operations are adversely affected. Therefore, sufficient personnel must be assigned to truck work (whether or not the crew is assigned to a separate truck company); the proper tools and apparatus must be available and the crew assigned should know how and when to use them; and the crew must be fully trained in all truck company fireground operations.

Fire fighters performing truck work should, then, have the training, equipment and manpower to carry out the five basic objectives of a fire fighting operation. Listed in the order in which they must be accomplished, these five objectives are:

- Rescue victims
- Protect exposures
- Confine the fire
- Extinguish the fire
- Overhaul the fireground

Salvage is sometimes considered to be a fire fighting objective rather than an operation. Although salvage does not contribute directly to control of a fire, it is very important in limiting the fire loss. In this book, salvage is considered a truck company duty and will be discussed, along with other truck company duties, in the chapters that follow.

Note that the list does not contain any of the "mechanical" movements, or evolutions, involved in fire fighting — such as raising ladders, advancing hoselines or ventilating. These operations are performed to accomplish one or more of the five basic objectives. For instance, truck crews might raise a ladder to allow a trapped person to climb down safely, but the objective is *rescue*, not the raising of the ladder. Fire fighting objectives are the same at every fire; a particular fireground situation dictates what movements are required to accomplish the objectives.

All but the last of these objectives are carried out in an atmosphere of flame and smoke. Thus, it is essential that fire fighters understand the nature of fire and the factors that affect its spread, including building construction, type of occupancy and types of fuel available to the fire.

FIRE SPREAD

Oxygen, fuel and heat are required to start and sustain the combustion process. These form the three sides of the familiar fire triangle. A more advanced concept of combustion includes a fourth element, a chain reaction phase, to form a fire tetrahedron. The chemistry of fire is not covered in this text, nor are the technical aspects of the support of combustion. Fire fighters are confronted with the problem after the fact. The discussions in this text are directed toward understanding how fire advances through a building and how it extends to exposures, since these are the characteristics that affect fire fighting operations.

In structural fire situations, the fuel and oxygen required to sustain a fire are generally in plentiful supply. The fire usually starts out small and, if attacked early enough, could easily be confined to the vicinity of its origin. When the fire burns unchecked, heat production increases. As the original fuel sources are consumed, the fire travels to new fuel sources in uninvolved parts of the building and in exposures.

There are three ways by which fire travels: convection, radiation and conduction.

Convection

Convection is the travel of heat through the motion of heated matter — that is, through the motion of smoke, hot air, heated gases and flying embers (Figure 1.1).

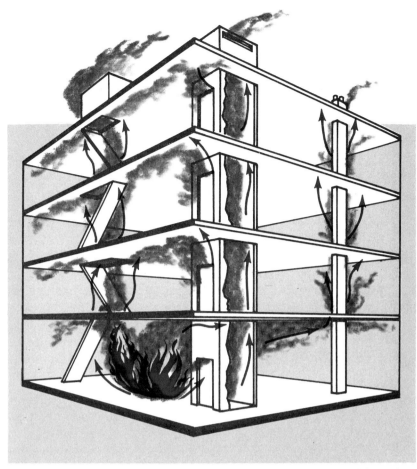

Figure 1.1. Convection carries hot air, smoke, gases and embers upward through available vertical channels.

When confined (as within a structure), convected heat moves in predictable patterns. The fire produces gases that, being lighter than air, rise toward the top of the building. Heated air also rises, as does the smoke produced by combustion. As these heated combustion products rise, cool air takes their place; the cool air is heated in turn and then also rises to the highest point it can reach. As the hot air and gases rise away from the fire, they begin to cool; as they do, they drop down to be reheated and rise again. This is the *convection cycle*. Within a building, this cycle will first fill the upper parts and then work down toward the fire.

It is easy to see how this method of heat and fire transmission creates a need for rescue operations and for checks of fire spread in the building. In addition, convection is the main reason for ventilation activities in fire department operations.

Modern aerial apparatus and required equipment have evolved over the years because of the convection problem in old brick, wood-joist buildings. Such structures, having open stairways, elevators, dumbwaiters and other shafts (many formed by their interior construction), permit rapid vertical spread of fire by convection.

Older building codes first passed in major cities did not require fire resistant construction, standpipes or other protection unless the building was over 75 feet high. This was probably because at the time these laws were passed the 75-foot aerial ladder was normally the longest ladder available.

Modern codes aimed at limiting fire spread by convection require that buildings of more than three stories be of fire resistant construction. Fire

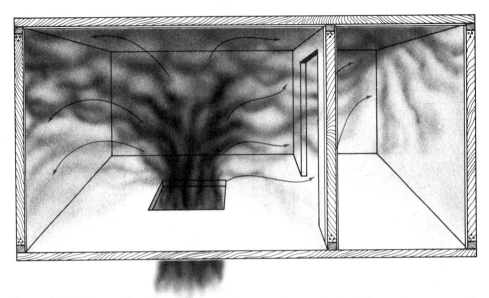

Figure 1.2. When vertical travel is blocked, convection carries hot air, smoke, gases and embers horizontally.

resistant construction is intended to confine the convection cycle to one floor or to a small area of a floor.

In any type of building, when fire is prevented from spreading upward, convection carries the fire outward. The gases spread across the ceiling and down walls and travel into adjacent rooms (Figure 1.2). Eventually the area is saturated with superheated gases. The intense heat quickly brings all combustible materials up to ignition temperatures, and the result is usually a violent production of flame that travels extremely fast from room to room.

Radiation

Radiation is the travel of heat through space; no material substance is required. Pure heat travels away from the fire area in the same way as light — that is, in straight lines. It is unaffected by wind and, unless blocked, radiates evenly in all directions (Figure 1.3). Once the fire has built to sizable proportions, radiation is the greatest cause of exposure fires, spreading fire rapidly from structure to structure or through storage areas. Intense radiant heat

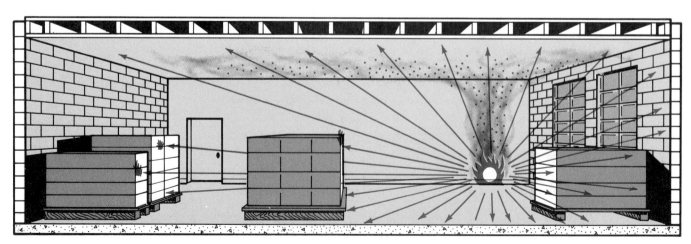

Figure 1.3. Heat is radiated evenly in all directions from the fire.

drives fire fighters back from normal approach distances, necessitating the use of heavy streams on the fire and exposed structures or stored materials.

In combination with convected heat, radiation creates the most severe area of exposure; this area must be protected first. However, although radiation is not affected by wind, the windward side of the fire cannot be ignored. Fire departments have been caught short too often when their efforts were directed solely to the area hit by both radiant and convected heat. After this most dangerous area has been covered, attention must be given to those areas exposed to radiant heat alone.

Within a building, radiant heat quickly raises the temperature of air and combustible material both near and, if the layout permits, at quite some distance from the fire. As a result, flashover can occur long before the flames actually contact the fuel in a given area (see the following section on flashover).

Proper ventilation is of little help against concentrations of radiant heat. Venting will remove the smoke, hot air and heated gases, thereby lessening the chance of rapid spread and flashover, but the radiant heat remains and must be counteracted through proper application of water on the seat of the fire.

Radiant heat, by itself and in combination with convected heat, can cause great physical distress to the fire fighter. For this reason it is imperative that full protective clothing be worn.

Conduction

Conduction is the travel of heat through a solid body (Figure 1.4). Although normally the least of the problems at a fire, the chance of fire travel by conduction should not be overlooked. Conduction can take heat through walls and floors by way of pipes, metal girders and joists, and can cause heat to pass through solid masonry walls.

If the spread of fire by conduction occurs at all, the time involved will depend on the amount of heat and fire being applied to a structural member or wall. In any case, when fire has been in contact with such parts, a thorough check must be performed to ensure that the fire has not traveled through them to other areas. Fire fighters must also be aware that heat can be

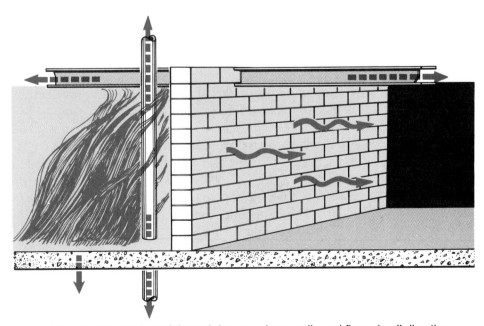

Figure 1.4. Heat is conducted through beams, pipes, walls and floors in all directions.

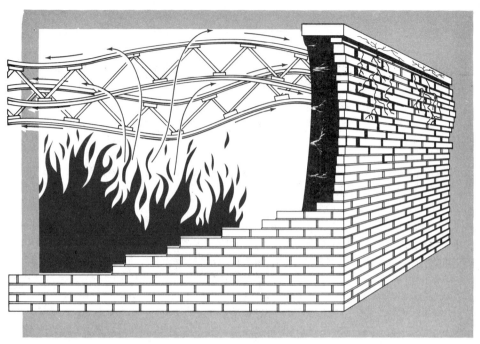

Figure 1.5. Conduction through exposed steel supports can cause them to expand, warp and possibly fail. This expansion might then cause walls to crack and perhaps fall.

conducted down, as well as in other directions, depending on the building design and features in the fire area.

Conduction also can be dangerous to fire fighters. Certain types of structures have steel building and roof supports that are completely open to fire. Heat spreading through these supports raises their temperature and can cause them to warp and fail, possibly causing the walls and roof to collapse (Figure 1.5).

In many cases, hose streams stop conduction without fire fighters being aware of it by removing heat from the structural members, walls and floors. Since heat traveling by conduction cannot be seen or felt in the normal course of fire fighting operations, the fire fighter must be observant and check for the spread of fire by conduction when there are indications that this could take place.

Flashover

Flashover is the ignition of combustibles in an area heated by convection, radiation or a combination of the two. The action can be a sudden ignition in a particular location, followed by rapid spread, or a "flash" of the entire area. The latter action is more likely in an open area within a building.

Convection can cause flashover at the top of a structure, owing to the hot products of combustion igniting materials at that level. Radiation can contribute to flashover in areas that do not block heat travel. However, flashover is not usually caused by radiation alone; radiation in combination with convection is more often a cause of flashover (Figure 1.6).

Smoldering Fire and Backdraft

The products of a fire can fill a building until the fire is almost starved for oxygen, at which point it will begin to smolder. The more incomplete the

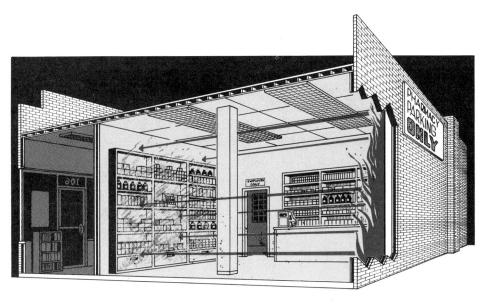

Figure 1.6. Flashover, heated gases and radiant heat traveling through a building can cause ignition at some distance from the fire.

combustion, the more carbon monoxide is produced. Carbon monoxide is a colorless, odorless, poisonous gas, especially dangerous because it is also explosive and flammable. Two of the elements necessary to produce fire — heat and fuel — are contained within the structure; only oxygen need be added. If oxygen is improperly allowed to enter the structure, the accumulated gases will ignite into a rapidly spreading fire or a violent explosion. This is backdraft. Fortunately, this situation can be controlled effectively through proper ventilation and attack procedures.

TRUCK COMPANY OPERATIONS

Truck companies are sometimes referred to as "ladder" companies, "hook-and-ladder" companies, "aerial" companies, and "snorkel" companies. Such labels might partially describe truck company apparatus, but they do not even hint at the planning, personnel, equipment and training that are coordinated in an efficiently operating truck company. A ladder truck and a driver do not make a truck company, any more than a pumper and a driver make an engine company.

Truck company apparatus and equipment have been designed to permit truck crews to function effectively and quickly in accomplishing the five fire fighting objectives. Through training and experience, truck company personnel must acquire knowledge, skill and judgment in performing the nine basic duties usually assigned to truck companies. These duties are:

- Rescue
- Ventilation
- Laddering
- Forcible entry
- Checking fire extension
- Salvage
- Ladder-pipe operation
- Utility control
- Overhaul

At some fires, it might be necessary for a truck company to perform all of these operations; other fires might require only some of the duties. Just as situations vary, procedures for each situation will also vary. With the exception of rescue, the duties are not necessarily performed in the order given above; that, too, depends on the fire situation.

For example, at a vacant building heavily involved in fire, the first truck duty might be to use an elevated stream. Other truck duties at such a fire might be limited to checking fire extension to adjoining or nearby buildings, laddering, and forcing entrance into these structures. On the other hand, a working fire in an occupied building could easily require that every truck duty be performed.

Knowledge of company territory, building inspections, and prefire planning are necessary if truck companies are to operate efficiently on the fireground, and are also useful in developing or improving on standard operating procedures. Truck company personnel should have a good knowledge of their first alarm territory, especially with regard to ventilation and laddering problems, life hazards, exposure hazards, locations especially dangerous to fire fighters, and other special conditions that affect fire fighting procedures. Perhaps it is impossible to learn everything about the company's area; but unusual situations should be carefully examined and analyzed, and special procedures should be developed when necessary.

This background information can be of particular help when the company arrives at a working fire and begins initial size-up. A continuing procedure, size-up results in operational changes to match changes in the fire situation. Factors to be considered in sizing up a fire situation, and the truck company operations required by different fire situations, will be covered in the remainder of this book.

SUMMARY

There are nine basic truck company duties. Each of these duties might or might not be required by a particular fire situation. However, truck companies should be staffed, equipped and trained to perform necessary duties quickly and efficiently, since their performance can affect the safety and performance of other responding companies. Truck company operations should take into account the ways in which fire, heat and smoke travel through a building, as well as the knowledge gained by prefire inspections and planning.

Truck duty can be one of the most challenging and rewarding assignments, provided that the truck companies in a department operate properly. This means that truck companies should be assigned to every building fire, that they should perform all necessary truck duties, and that they should not be made to function simply as personnel carriers — to bring more hose crews to the scene.

Truck companies must not be considered solely for the use of their aerial units, or worse, only for elevated streams. Departments that relegate truck companies to these very limited operations do themselves and the community a disservice.

At most fires, the outcome depends on the efficiency with which first alarm operations are carried out. The efficiency of truck company operations is a most important factor in determining the overall fireground efficiency.

INITIAL ASSIGNMENTS 2

Most fire departments — paid or volunteer — must operate with a limited number of personnel. If a truck company is to be ready for any fireground situation, it must be set up so that its personnel can perform all the required duties in as short a time as possible. Response patterns and operating procedures must ensure that crew members get into action quickly and that the most urgent operations are begun immediately upon arrival on the fireground.

For this, each crew member must be trained in, and thoroughly familiar with, the use of all the equipment on the apparatus. Before the truck arrives on the fireground, all crew members should know which tools they will be carrying into the fire building and what their duties will be on arrival — that is, their initial assignments. Standard operating procedures should detail which parts of a fire building each truck company is to cover and how the apparatus is to be positioned, given the particular fire situation. In other words, both the company as a whole and the individual crew members must know their initial assignments before they arrive on the fireground.

This chapter deals with the tools carried on the truck, the assigning of responsibility for these tools to crew members, coverage of the fire building, and apparatus positioning. Included in the discussions are some simple but effective operating procedures to ensure that all truck crew members are aware of their initial assignments and able to accomplish them. However, it is important to realize that all fire fighting procedures require training if they are to be performed efficiently in a fire situation.

TOOLS AND PERSONNEL

The truck is designed to carry ground ladders, tools and, almost always, an aerial unit. The sizes and types of ladders carried on a truck will depend on minimum standard recommendations and company and department experience. Crew members must be trained to handle and raise ground ladders and must be ready to place them when they are called for by the truck officer. The aerial unit should be ready for service as needed. Aerial operations and ground laddering are discussed in detail in Chapters 8 and 9.

The tools are the most-used items on the truck. They are employed at almost every fire, for several of the truck company duties listed in Chapter 1.

Hand Tools

Hand tools most useful for forcible entry are the claw tool, the kelly tool, the halligan tool (Quic-Bar), and lock pullers such as the K tool and the flathead axe. These tools can be used to break locks, to force exterior and interior doors, to open sidewalk doors and grates, and generally to provide entry to a building. The flathead axe also can be used to drive other tools, including a second flathead axe.

For ventilation work, the pickhead axe, pike pole and halligan tool are usually preferred. The axe can be used to cut and rip open roofs, to force or knock out windows, and to cut and force natural or built-in openings such as skylights, ventilators and roof hatches or scuttles. The halligan tool is used to rip up cut roofing, to force and knock out windows, and to force natural openings.

The pike pole (ceiling hook) can be used in ventilation work to pull down a ceiling or to push the ceiling down from the roof after the roof has been opened. It is also useful for opening a transom below a skylight, removing coverings below roof hatches, and knocking out window glass.

All these tools can also be used in checking for fire extension and in overhaul. For these operations, they are used to pull down ceilings and walls, rip up floors and baseboards, and generally open the interior of a building for inspection.

Power Tools

Power tools can be of great help in truck company operations, especially where lack of personnel is a problem. The tools can be powered by electricity, gasoline engines, or air or hydraulic pressure. Although they can ease and speed up the work of the truck company, they have one major disadvantage: except for some hydraulic spreading devices, power tools can only be used for cutting. They cannot be used for prying, breaking or just plain forcing into and through a building. They are also larger and heavier than hand tools and can take more time to get into position. However, they will cut much faster than hand tools.

Although power tools cannot replace hand tools, they do have a place in truck company operations and should be given careful consideration when equipment is being purchased.

Training

Truck crews must be thoroughly familiar with the tools of their trade. This means that they must be trained in the use of hand and power tools as well as in the handling of ladders. Most truck companies can find places in their territory where they can raise ground ladders, position the aerial unit for rescue or ladder-pipe operations, and carry out other training movements. However, truck company tools are used to break things, so training usually involves the destruction of property. Opportunities for complete training are thus quite rare, since no owners want the doors of their buildings forced open, their roofs cut up, or their windows knocked out simply for training purposes.

Since the weakest part of the continuing training program will be tool use, fire departments, truck companies and truck officers must take advantage of every training opportunity. Buildings that are about to be torn down can, with permission, be utilized as tool-use training structures. Companies might also be allowed to use their tools in parts of a building about to be remodeled. Mockups of buildings have limited training use, but they are better than nothing. The best training structure is one in which truck crews can employ all their tools as they would in an actual fire situation, so they are prepared to perform all necessary fireground operations.

Tool Assignments

The truck officer should not have to tell each member of the truck crew which tools to take into the fire building or exposed buildings. Instead, as a part of prefire planning, personnel should be assigned to particular tools based on the locations of the tools on the truck and the positions taken by fire fighters as they board the truck.

Hand tools and power tools should all be located on the truck so they are readily available. Such hand tools as axes, claw tools, halligan tools, pike poles, lock pullers and kelly tools should be located at or near the positions in which the crew will ride to the fire. Assignment of a crew member to a particular truck position should then include responsibility for the tools near that position. The fire fighter should take these tools into the building, either immediately upon arrival or after performing necessary outside duties.

The truck officer must ensure that the tools are properly located on the truck and must assign crew members to positions, or to tools, in such a way that the proper assortment of tools is taken into the building.

In paid departments, fire fighters should be assigned truck positions and tools when their shift assumes duty. In some departments, all members are assigned positions on the truck, and this determines the tools for which they are responsible. In other departments, all members are assigned groups of tools, and this determines the positions they will take on the truck. Either system is effective if followed consistently.

In some city departments, one member of a truck company is assigned to the "bar" or "irons" and works with engine company personnel upon arrival at the fire (Figure 2.1). This fire fighter is equipped with a claw tool, halligan tool or similar device, a flathead axe, self-contained breathing apparatus, and the usual running gear, and might also be equipped with a small porta-power or a rabbit tool. This crew member forces entrance for the engine crew, ventilates the fire floor around them, and searches for victims in the fire area and — where the situation warrants it — above the fire.

Note that the "bar" or "irons" person belongs to a truck, rather than an engine, company. If transferred to an engine company, this fire fighter would soon be advancing hoselines, and not performing required truck duties.

In volunteer departments, tool assignments should be based on the positions taken by crew members on the apparatus. A chart of the truck showing crew positions and tool locations can be helpful in making these assignments. The chart should be kept on display in the station and made a part of the training program (Figure 2.2). It may include an "irons" person to be assigned to work with engine personnel, as noted above.

Volunteers who come to the station when there is no fire alarm should, upon arrival, be assigned truck positions and tools according to the chart. Top-priority positions should be assigned first. This immediate assignment eliminates the confusion that might result if several crew members attempted to

Figure 2.1. Fire fighter assigned to "bar" or "irons" ready to enter building. Note small porta-power unit.

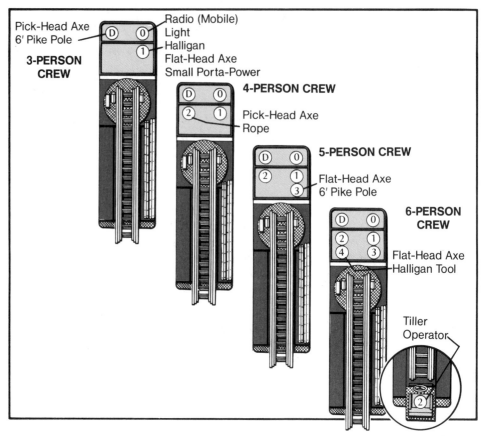

Figure 2.2. Examples of truck company personnel positioning and tool assignments.

take the same position when an alarm is received. Volunteers who arrive at the fire after the truck is positioned should check to see that the necessary tools have been removed and taken into the building.

COVERAGE

A properly equipped, staffed and trained truck company should be capable of operating efficiently from the moment it arrives at a fire. This means that the company cannot wait until it is on the fireground to decide on an initial course of action. The crew must know, before they arrive, what their movements will be. For this, it is necessary to develop a standard response procedure.

The details of any standard procedure will vary according to the types of structures and occupancies in a territory. The procedure will also be affected by the number of companies responding to a first alarm, the distance between their stations, and the normal personnel complement of each responding company. The manner in which the procedure is implemented will depend on the construction and occupancy of the involved building and exposed buildings.

A standard response procedure allows responding companies to begin operations as soon as they arrive on the fireground, since they are already aware of their initial assignments. Moreover, each company — and each fire fighter — knows what the others are doing since all personnel are operating according to the same plan.

Front and Rear Coverage

To be effective, a standard response procedure must provide for immediate front and rear coverage of the fire building by initial alarm units. Then the

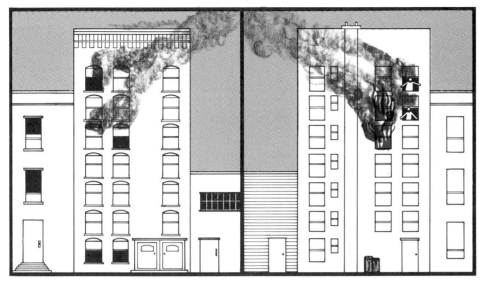

Figure 2.3. Fire conditions may vary from the front (left) to the rear (right) of an involved structure.

sides of the building (if detached) should be covered. Fire conditions often vary between the front and the back or sides of a building; it is important that the entire building be checked as soon as possible upon arrival (Figure 2.3).

If two truck companies are responding to an alarm and they are not far apart, the truck that is expected to arrive first (the first due) should be assigned to cover the front of the building. The second truck is assigned to the rear. These assignments should be modified according to the situation: if, upon arrival, the first due company finds a life hazard or some other serious situation in the rear, it should cover that position first; the second due truck should be notified by radio of the change in procedure and advised to cover the front.

The assignment of the second due company to rear coverage does not mean that the truck must be driven to the rear. In some cases, this is impossible. It does mean that crew members must check the rear to determine what the situation is with regard to possible victims and the extent and intensity of the

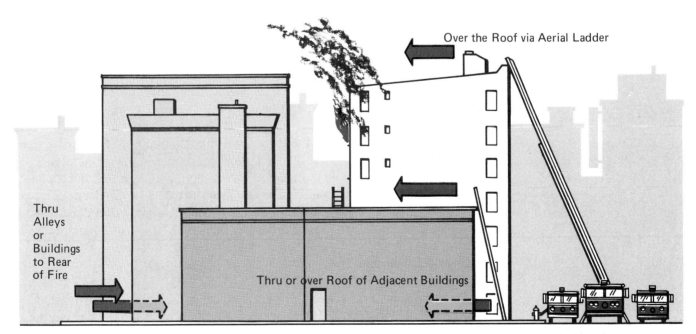

Figure 2.4. For effective building coverage, truck crews must use various paths to get apparatus or personnel to the rear.

fire. They should check the locations of exposures and of rear stairs, porches and basement entrances that might be used in fire fighting operations. They should determine whether assistance is required by engine company personnel working in the rear. If necessary, they should provide ground ladders, to be used to advance attack lines.

If only one truck company responds to the first alarm, at least one person from the unit should be assigned to check out the rear immediately upon arrival.

To reach the rear of a fire building, truck company personnel might have to move through adjoining buildings, through the lower levels of the fire building, or over row buildings (Figure 2.4). Any alleys, walkways, courts and areaways can be used to gain access to the rear of a building. In some cases, it may be best to position the apparatus on the street at the rear of the fire building for both rear coverage and truck company operations. The position of the truck has much to do with the efficiency of the coverage operation (positioning is discussed in the next section). For maximum efficiency, the company should get to know the area in which it operates, through prefire inspection and planning.

Truck Out of Quarters

Standard operating procedures that assign coverage responsibilities to first alarm truck companies must assume that these companies will respond from quarters. Occasionally, a truck company will receive an alarm by radio when it is completely out of position, perhaps in one corner of its territory. The company should advise the dispatcher of the situation by radio as soon as the alarm is received. The dispatcher can then either send a closer truck from quarters or advise the second due truck that it will be the first to arrive.

Other Aspects

Some departments assign responding companies to interior coverage as part of their standard response procedures. For example, the first due truck might be assigned to cover the fire floor as well as the front of the building. The second due truck then covers the floors above the fire and the roof, along with the rear of the building. The effectiveness of such variations in the front-and-rear coverage pattern depends on the particular department's fire problems (Figure 2.5).

Truck companies do not work alone at a fire. The standard response procedure must apply to responding engine companies as well as to truck companies. The coverage operations of both types of companies must be coordinated so that the companies responding to a first alarm work as a team with minimal duplication of effort. The important point is that the fire building must be covered quickly, front and rear, outside and inside. For fire fighter safety, vacant or abandoned buildings should not be entered if a major fire is encountered. Coverage is the responsibility of all first alarm companies, and it must be performed according to a standard procedure. Chief officers should not have to give orders to arriving first alarm companies.

Size-up Information

The objective of the coverage operation is to gather enough information to size up the fire situation accurately. Crew members covering the fire building

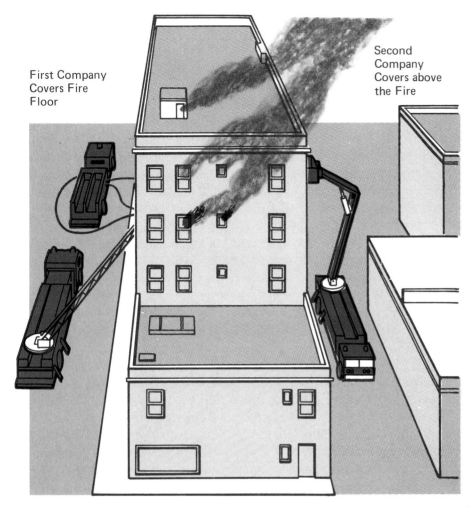

Figure 2.5. Interior coverage, which can be part of a standard operating procedure, should be based on a department's specific fire problems.

should report their information promptly to the company officer, or to the chief officer if that person is more readily accessible. Where possible, reporting by radio will speed up the delivery of information. The information should be precise and accurate. Negative situations (such as when a position cannot be reached or when requested information cannot be obtained) should also be reported. With reports from all areas, the officer-in-charge should quickly develop a good mental picture of the situation. The fire fighter's size-up will determine which operations should be initiated next, and whether or not to call for additional companies.

Both the fire situation and the number of truck personnel at the scene affect the decision on which operations to initiate. First priority must be given to operations whose objective is rescue; second, to exposure protection; and so on through the list in Chapter 1, in the given order. Salvage operations should be performed where possible to limit water damage. The smaller the crew, the more important it is that duties be assigned according to these priorities.

The number of personnel available determines how much can be done at one time. Five or six fire fighters might simultaneously conduct interior search and ventilate when both operations are required for rescue, whereas a three-member crew might only be able to perform one of these operations at a time. Which operation is performed will be determined by the size-up. Although it is most desirable to enter the building and search for victims, the building might have to be ventilated before it can be entered. In some cases, ladders will have to be raised to ventilate the building, or forcible entry may be required so personnel can get in to search.

Because truck operations affect the entire fire fighting effort, the officer-in-charge should not hesitate to call in additional companies as needed. It is extremely important that all necessary truck operations be completed as quickly as possible. These operations should be performed simultaneously, rather than one after the other by the same tiring crew.

If in certain areas of a company territory more than one truck crew will obviously be required, then two or more truck companies should respond to first alarms in those areas. The additional companies can be part of the department in whose territory the fire occurs, or they can come from neighboring departments. The important thing is that they be on their way to the fire at the first alarm. If size-up indicates that they will not be needed, the first due company officer can let them know by radio, and they can return to service; if they are needed, they will arrive on the fireground in time to help save lives and avert undue property losses.

APPARATUS POSITIONING

For most effective operation, the truck company apparatus must be positioned properly on the fireground. This is also true of the engine company apparatus, when a hydrant-to-fire approach is used for fire attack. However, there need not be a conflict between the two units; trucks and pumpers alike can be positioned for effective operation with a little foresight and planning. As with coverage assignments, standard procedures can make apparatus positioning a smooth and efficient operation.

Prefire plans should not require that the truck be positioned in a certain way at every fire or that truck crews perform certain duties regardless of fire conditions. Nor should the truck be positioned so that the aerial unit can be raised whether it is needed or not. Such rules, including stipulation that the aerial unit should be raised at every fire for training purposes, are ridiculous. They lead to poor positioning, unnecessary operations and a waste of time, effort and personnel. Training should be done at training sessions. Truck companies must be positioned and assigned as each fire situation dictates.

Approach

A ladder truck is not the easiest vehicle to maneuver in tight quarters or in close proximity to other apparatus. The closer the truck is brought to the fire building, the more difficult it is to move. The approach to the fire building is therefore very important.

Once the truck is at the fire scene, it should be advanced in a slow, deliberate approach to the fire building. No company should try to beat another company to a position; such foolishness can only lead to poor positioning, accidents and injuries. The officer-in-charge should be concerned only with getting a good position from which the company can work efficiently. The officer's approach to this position must be made in accordance with standard operating procedures and the fire situation.

The truck must not be committed (that is, stopped, the wheels chocked, the jacks set, etc.) until it is in the proper position. This can only be determined after the truck has made its approach to the fireground and the fire conditions have been observed.

Positioning

In areas of one-story and two-story buildings, it is not usually essential that the truck be positioned directly at the front (or rear) of the fire building. Since the aerial ladder will probably not be used in truck operations in such low-rise structures, the truck can be positioned to one side of the building. This will leave the priority positions for the engine companies but will not hinder truck operations. Truck crews will be able to quickly remove ground ladders and special equipment from the truck as required.

A truck should, however, avoid a position that blocks engine companies from a hydrant or protective system intake.

If the fire building is more than two stories high, the engine companies must allow the truck companies to get close to the building so aerial units can be positioned effectively and the larger ground ladders can be raised quickly. For example, suppose an engine company and a truck company are approaching the front or rear of a fire building from the same direction, with the pumper ahead of the truck. The engine company should pull on past the building to allow the truck to be positioned properly. If the sides of the building are detached, this will also allow the engine officer to see three sides of the structure. The engine officer should report observation of the far side to the truck officer, who will have seen only two sides (Figure 2.6).

If an engine company and a truck company are approaching a taller building from opposite sides, the pumper should stop short of the building. This, again, will allow the truck company to get its apparatus into a good position. In this case, each officer will have seen two sides of the building (if the sides are exposed); they should exchange information quickly, as operations begin.

In some cases, the width of the building and the location of the fire within the building may solve the problem of positioning the trucks and pumpers. For instance, at a fire in a wide apartment or commercial building, the engine would be positioned near an entrance to allow engine crews to advance into the building and control the fire in stairways and corridors. The truck should be positioned for rescue, ventilation or other truck company operations (Figure 2.7).

On the other hand, the extent, location and intensity of the fire might not allow for perfect positioning of the truck. In these cases, the truck should be

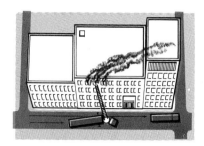

Engine and Truck Responding from Same Direction

Engine and Truck Responding from Opposite Direction

Figure 2.6. For front coverage of an attached structure, the first-arriving engine should be positioned according to the approach of the first-arriving truck.

Figure 2.7. When wide buildings are involved, the truck company must be positioned to accomplish the most important operations required of the aerial equipment.

Figure 2.8. When row structures are involved, trucks can be positioned at adjoining buildings to get fire fighters to the roof. This should not be done if the aerial ladder is required for rescue at the fire building.

positioned as well as possible and operations initiated without delay. If the building height and the intensity of the fire require such action, the truck should be positioned for use of the ladder pipe or platform pipe. Or, the best plan of action might be to position the truck some distance from the building to allow fire fighters to perform their duties without unnecessary hazards.

When the fire building is one of several row buildings of the same height, the truck can be positioned so the roof of an adjoining building can be laddered. The fire fighters can then work their way from the adjoining roof to the fire building roof for venting and to the rear for other duties (Figure 2.8).

SUMMARY

Truck duties are performed at every fire. A truck company is outfitted with the tools and equipment needed to carry out these duties, and no truck crew member should leave the truck empty-handed. Procedures should be devised to assign responsibility for truck tools and ensure proper coverage and apparatus positioning at the scene of a fire. Proper coverage means, at the least, covering and reporting on the situation at the front and rear of a fire building. Proper positioning includes the positioning of trucks as well as pumpers for most efficient fireground operations.

RESCUE 3

Every fire fighter knows that the rescue of people in danger is the primary objective of a fire company and the first duty to be performed at the scene of a fire. However, not every fire fighter realizes that rescue duties involve much more than the bodily removal of people who might be trapped. Although carrying a fire victim to safety is rescue in the purest sense, rescue work includes many other operations.

Raising a ladder for use by entrapped occupants, assisting or directing people from the fire building, and searching for victims in the building are all rescue operations, each of which immediately reduces danger to human life. Rapid ventilation removes accumulations of smoke and gases and prevents their further buildup. Proper placement of the first hoselines at a fire can keep the fire away from people in the building. Both operations reduce the danger to entrapped occupants and extend their time to get out of the building. In a very real sense these, too, are rescue operations.

Many fire fighters tend to think of rescue only in connection with hospitals, nursing homes, schools, hotels and other occupancies containing many people. Such buildings must receive careful consideration in terms of rescue problems because of the number of people involved, but fires in one- and two-family dwellings very often require rescue operations. A review of national statistics shows that injuries and deaths in dwelling fires far outnumber those in other occupancies.

For truck companies, rescue is a complex operation. Almost every rescue situation calls for a different combination of movements, equipment and other operations. These can include forcible entry, interior search, placement of ground ladders, and ventilation.

This chapter discusses rescue work, beginning with prefire planning; search is considered in greater detail. The uses of ground ladders, aerial units, and forcible entry and ventilation procedures for rescue are discussed in later chapters.

CHRONOLOGY OF RESCUE OPERATIONS

Before the Alarm

Preparation for rescue begins well before the alarm is received. It begins with building and area inspections and continuing study of the company's territory to determine the occupancies, the people involved, the hazards, the potential rescue problems, and the most effective apparatus positions. The objective is to know in advance the approximate type and extent of rescue operations that might be required at any fire.

At the Alarm

The information received with the alarm can be the first indication that a rescue problem exists. The initial information might include an exact address or a more general location such as a street intersection. From this information, truck company personnel should know the type of occupancies involved or the type of area to which they are responding. This knowledge and the time of day are two very important clues to the possible presence of victims in the fire building and to the type of rescue problems likely to be involved (Figure 3.1).

The information given with a verbal dispatch can be important. Such phrases as "across from," "next door to," "at the rear of" and "near the intersection of" can indicate that the alarm was not turned in by an occupant of the building, but by someone outside who saw smoke or flame coming from the building. People inside the building might be unaware of the fire or unable to escape it at the time of the alarm.

Figure 3.1. The type of occupancy and the time of day are the two most important clues related to rescue operations.

First reports, whether received by a central dispatcher or by someone on watch, should be relayed *in their entirety* to the company officer. This will allow the responding companies to extract as much information as possible from the reports, including indications of any rescue problems.

On the Fireground

Size-up. Even before the truck is stopped, the officer should have begun a careful size-up of the situation. Answers to the following questions will help to determine the extent of rescue operations.

- Is the fire building a closed-up dwelling with much heavy smoke showing?
- Are cars parked in the driveway, front, or rear, indicating that an entire family might be inside?
- Are people at the windows of an apartment house, office building or similar occupancy, calling for help?
- In such a multiple occupancy, with smoke showing, can calls for help be heard coming from the inside?
- Is fire showing? Where?
- Given the interior construction of the building, in which directions will the fire travel most rapidly?

In addition, the extent of the fire, the size and age of the building, and its apparent population are important in ascertaining what rescue operations are needed. Some of this information would have been obtained in prefire inspections.

Other size-up information can be obtained from neighbors and from tenants who have escaped the fire building. Of special urgency are reports that people are still inside. On the other hand, reports that "everyone is out" might be erroneous and should not deter fire fighters from starting search operations, especially in multiple-family residences. The size-up will indicate where search and rescue operations should begin.

Immediate rescue. Immediate rescue, at the expense of other truck company operations, must be attempted in extreme cases — such as when an arriving company finds occupants about to leave the building by jumping or occupants whose clothing has caught fire. In such situations, truck crews must delay all other operations in favor of raising ladders. They must get the attention of victims to make it obvious that rescue is under way, and must talk to the victims to calm them until they can be brought down. Battery-powered megaphones are useful for this purpose.

The presence of victims at windows, needing help to get out, often indicates that other occupants are unable even to reach windows and are trapped within the building. When any single rescue operation keeps fire fighters from other duties, help should be called for immediately.

Often, the sight of fire fighters arriving on the scene tends to calm panicky people. Otherwise, fire fighters must take immediate action to control overexcited occupants. One way to do this is to give *positive* orders and directions, in a forceful manner. A weak or negative approach, as in "Okay, folks, just take it easy," or "Let's not panic," will produce negative results and should *never* be used. Firm, positive orders, like "Move back," and "Use the stairway behind you," will have a calming effect on victims trapped in the building and will give fire fighters a chance to rescue them. This also holds for persons encountered during search operations.

Obtaining water and placing streams. This is the function of responding engine companies. During rescue operations, streams are used to

- Separate the fire from the people closest to it
- Control interior stairways and corridors for evacuating occupants and advancing fire fighters
- Protect truck crews searching for victims around and above the fire

Streams should be placed as soon as possible upon arrival to coordinate with the search.

Search. If there is any indication that victims might be trapped or overcome within the fire building, a search should begin immediately. (Search is discussed in detail in the last section of this chapter.) Perhaps no other operation demands as much cooperation and coordination between truck and engine companies. If truck crews are to search around and above the fire, they must know that the fire is under attack and that attempts are being made to block fire spread. To use their lines effectively, engine companies might require truck operations, such as ventilation, laddering and forcible entry. These operations may also be required before the search can begin.

It is extremely important that every fire fighter at the scene be aware that a search is in progress. All activity should be directed toward helping truck crews engaged in the search, and providing protection for them and for any victims they may find.

Ventilation. The building should be ventilated as soon as possible to allow smoke and gases to move away from any occupants who could be trapped.

RESCUE CONSIDERATIONS

Rescue means removing victims and potential victims from danger. The extent of the rescue problem is directly affected by

- The number of people in the fire building
- The paths by which fire and smoke can reach them
- The routes available to truck crews for reaching people and removing them from the building

These factors, in turn, depend on the construction, size and interior layout of the building.

This section deals with rescue problems in several types of occupancies. It is important that the truck company be prepared — equipped, staffed and trained — for efficient rescue operations in the most complex structure in its territory. As an extreme example, if an area contains only single-family dwellings, with the exception of one nursing home, truck companies in that area must be as well prepared for rescue operations at the nursing home as they are for operations at the dwellings.

At any fire, but more often in larger occupancies, the rescue problem can be great enough to tax the capacity of the first units at the scene. If it is apparent or even suspected that such a rescue problem exists, additional companies should be called without delay.

Figure 3.2. During size-up, truck crews must look for signs that people are in the building.

Residential Occupancies

Fires in residences of any type might require rescue operations at any time of day. At night, the rescue situation is usually more acute since occupants are asleep and off guard. In addition, there are usually more people in a residence at night than during the day (Figure 3.2).

Single-family dwellings. In a typical two-story dwelling with two or three rooms on fire on the first floor, the occupants in most danger are those close to the fire on the first floor or those directly over the fire on the second floor. The former will be affected by the radiant heat produced, and the latter by the convected smoke, hot air and gases.

The main body of fire should be attacked immediately. As the attack on the fire progresses on the first floor, the area around the fire should be searched thoroughly. At the same time, truck crews should be sent to the area over the fire to begin ventilating and searching for possible victims there. No time should be wasted in getting fire fighters to the upper floor. If an attack line is immediately available, it should be taken upstairs. If not, search should begin without it.

If fire is discovered in any upstairs room, that room should be searched if at all possible. Then the door should be shut to isolate the room until a line can be advanced. The doors and windows of other rooms should be opened to provide ventilation, dissipate heat and smoke, and allow more efficient search and fire attack operations.

In addition to the front entrance and stairs, access to the second floor can be by rear stairs, porches and ladders. Search of the bedrooms and other truck duties can be carried out while engine company personnel advance lines on the floor below to control the stairway.

Apartment houses. In any large, occupied building the location of the fire and of most of the smoke above the fire should be carefully noted during size-up. Smoke indicates the area into which the fire will most likely spread, the path it will take, and the location of occupants who will be most endangered if the fire does spread (Figure 3.3). More victims are overcome by carbon

Figure 3.3. In a large, occupied apartment building, the location of the fire and the direction of the most smoke should be carefully noted, since those areas pose the greatest danger for occupants. For example, the victim at the upper right window may be in greater danger than the victim at the lower left window, even though the victim at the right is further from the fire.

monoxide than are burned. Therefore, while the fire is being attacked, every effort should be made to vent the building.

A search of the fire floor, the floor above the fire, the top floor, and then other floors should be started as soon as possible to make sure that all occupants are located and removed from exposure to toxic gases. Here, even more than in a dwelling, search and rescue must be coordinated with a properly mounted attack on the fire.

Hotels and motels. These residential occupancies present problems similar in some respects to those of large apartment houses. The rescue problems depend on the size, age, general construction and population of the building. However, although there are usually fewer people per unit in a hotel or motel than in an apartment building or rooming house, with buildings of similar size there will be more units in the transient type occupancy. Thus, as many people can be found in a hotel or motel as in an apartment building of the same size.

The number of people in a transient occupancy varies with the time of day and day of the week. Hotels, and especially motels, contain many more people at night.

When a working fire is encountered in this type of building, it must be thoroughly ventilated. The fire must be attacked, and the fire floor, the floor above the fire, the top floor, and then the other floors must be searched. In the typical two-story to four-story motel, the rooms directly behind those on fire must receive prime attention because of the possibility of lateral travel of smoke, gases and fire between units.

Industrial Occupancies

Factories, warehouses and other industrial buildings present the greatest rescue problems during the usual daytime working hours. However, in many instances, a second shift works well into the night, and a third shift is employed for a 24-hour operation. To be properly prepared for all rescue situations, truck companies assigned to industrial areas should be aware of the working hours in all such occupancies.

Another rescue consideration is the general physical ability of the employees. For example, the rescue problems in a factory employing mostly women will differ from those in a plant staffed mainly by men. Still other rescue problems can occur in a factory where a special effort is made to hire handicapped workers.

Rescue can be hampered or complicated by burning chemicals or other hazardous materials. Large areas such as warehouses often require special rescue procedures to maintain control and contact with fire fighters engaged in search.

Such variables affect the success of a potential rescue operation and point to the need for (and value of) prefire inspection and planning.

Since offices are usually empty at night, daytime fires present the major rescue problem in these occupancies.

Hospitals, Schools and Institutions

Fires in hospitals, schools, nursing homes and similar institutions are handled in essentially the same way as fires in multiple-family housing. Here, however, the search and rescue problem is compounded by the perhaps larger number of people and by their age and physical condition.

Although at night a hospital usually will have a smaller staff, no visitors, and a reduced maintenance crew, the smaller number of people does not make night rescue easier. Sleeping patients, sometimes sedated, often immobile, will be without the benefit of a full staff to assist them to safety. If the fire is in a vacant work area (such as a kitchen, storeroom or workshop), it might not be detected promptly by a small night staff; this could delay the alarm and increase the severity of the rescue problem. The same situation can be compounded in nursing homes by the high ratio of elderly occupants.

Special areas are often reserved for bedfast, nonambulatory patients, and for patients who require continuing care. Fire companies must know the locations of such areas within hospitals and nursing homes in their territory since these patients require extraordinary rescue procedures. Whether or not

Figure 3.4. According to the conditions, patients can be lowered to the floor or moved to an area clear of smoke.

it is necessary to evacuate such an occupancy depends on its construction and size as well as the location and severity of the fire. If it is evident that the fire will be controlled, it could be possible to move patients within the building to areas that are isolated from the fire. This is true especially where corridors are divided by smoke barriers or fire barriers with appropriate smoke and fire control doors. Patient room doors are usually wide enough to accommodate a hospital bed or a mattress used as a stretcher to permit such movement.

When smoke is the main problem, it might only be necessary to lower patients to the floor to get them below the smoke level. This can be done by lifting the mattress, with the patient on it, off the bed and setting it on the floor.

Either of these actions might be desirable in view of the physical condition of the patients, the continuing care some must receive, and possible adverse weather conditions. However, if there is any doubt regarding control of the fire, the building must be evacuated completely (Figure 3.4).

Most schools present a rescue problem in the daytime on weekdays from September to June. Nowadays, however, schools are often used at night and throughout the year for adult education, athletics, extracurricular activities, and so on. It is the responsibility of every truck company to be aware of such uses of the schools within its first alarm response area.

Upon arrival at a working fire in a school, fire fighters should expect to find overly excited or panicky children and adults. The school should be completely evacuated. Areas around and above the fire should be searched immediately if it is not certain that everyone has left the building.

Retail Stores

Stores are, for the most part, occupied during shopping hours and empty at other times, but during certain selling seasons shopping hours are extended and stores are often overcrowded. Immediately before Christmas, for example, most stores are crowded with both people and stock. This is also true, to a lesser extent, before Easter, during school holidays, and before school starts in September. In all seasons, retail stores are most crowded on weekends and in the evenings.

Thus, the time of day, the day of the week, and the season all affect mercantile rescue operations. In addition, it should be noted that the occupants — both customers and employees — of most shops are women. Compounding normal rescue problems is the fact that many young children accompany the customers. The children are often poorly supervised and tend to roam through the store. A report of fire can quickly lead to mild panic as some adults search for their children and others rush for the exits.

Exits. In general, the larger a store, the greater the distance between most customers and the exits. This makes it comparatively difficult for customers in large stores to get to an exit. Modern one-story "super stores," such as supermarkets, huge drug stores, and discount stores, are laid out for restricted traffic flow. Narrow aisles between merchandise displays lead to even narrower checkout aisles and finally to only one or two exit doors. Entrance doors might be blocked by turnstiles designed only to let people into the store, not out. Shopping carts often are stored where they increase the difficulty of evacuating the building.

Door openings are usually limited in size in relation to the floor area of the store and the number of people the store can hold. With fire and smoke closing in on customers, arriving fire fighters could find a panicky group of people pushing and shoving but unable to get out of the building. The main fire department operation here is to get the people out. Truck companies must evacuate the occupants and assist engine companies in getting into the store with attack lines.

Additional openings. Truck crews must calm the occupants and establish a traffic pattern out of the store. They must ensure that occupants move only in one direction, so they will not clash with each other or get in the way of entering engine crews. Revolving doors can be jammed by people pushing on both sides; truck crews must trip (collapse) the doors to allow occupants to get out. If necessary, additional openings must be forced to allow occupants to leave and engine crews to enter.

In most modern super stores, storefront windows open directly to the sales floor. There is usually only 1 to 2 feet of exterior wall below the window, so, with the glass removed, these window openings can be used as additional exits for customers and entrances for engine crews. If possible, two separate windows should be removed. Customers should be directed to one opening, and engine crews should use the other to avoid congestion.

Figure 3.5. Store windows can be used for moving out occupants and moving in fire fighters.

A plate-glass window must be removed carefully, so people in the store are not showered with broken glass. (If possible, a truck crew member should work from inside the store to knock out the glass toward the outside.) The glass is first broken and knocked out at the *top* of the window with a pike pole, then the rest of the window is cleared of glass from the top down. Fire fighters should now enter the store and direct occupants to the opened window, away from the opening made for engine company personnel (Figure 3.5).

Search. Following the evacuation (or at the same time, if personnel are available) aisles, offices, and storerooms must be searched. Occupants might have sought shelter in these places or might have collapsed or fallen on the way out. As with multistoried residences, the upper floors of multistoried stores must also be searched. The search is generally simplified by the lack of interior partitions.

Evacuation and search procedures can be carried out most efficiently by truck crews who, through prefire inspection, have come to know the construction and layout of the stores in their territory. As noted above, the locations of windows and the operating mechanisms of entrance and exit doors are important in rescue situations.

SEARCH

A thorough, planned search for victims should be conducted at every fire. Moreover, all fire fighters, no matter what type of company to which they are assigned, should be able to conduct a search if the need arises. All personnel should realize that the safety of fire fighters engaged in search is their responsibility. Crews operating a line on the fire floor must be aware that truck crews will be searching the floors above for possible victims. If those on the line cannot control the fire and are forced to retreat from the building, truck crews above must be warned so they can take appropriate action. Fire fighters assigned to laddering duties must also be aware that a search is in progress, since they might have to place ladders as additional exits for search personnel and victims.

Since search is an overall responsibility, it should be performed according to an efficient standard procedure that is safe for all fire fighters. The procedure should be simple and straightforward so that one fire fighter can substitute for another at any point in the search. It should include an easy-to-follow search pattern that is sure to begin where there is most danger to occupants.

Normally, occupants closest to the fire are in the most danger, whether they are on the fire floor or the floor above. Those on the fire floor will be affected by radiant heat, smoke and gases, and those above by convected heat, gases, smoke and hot air. Occupants who are two or more floors above the fire (and especially on the top floor) can be endangered by smoke and heat that have been channeled up through the building. Therefore, truck crews must begin both search and ventilation operations quickly, while engine crews attack the fire.

Search Duties

It is important that a number of operations be carried out simultaneously — or as nearly so as possible — in any rescue situation. While following a standard search procedure, truck crews should perform the following duties, the first being by far the most important:

- Locate and remove trapped occupants
- Ventilate where needed
- Temporarily prevent extension of fire by closing doors and windows
- Check for interior and exterior fire extension
- When necessary, help locate the seat of the fire

Standard Search Procedure

The best way to see the value of a standard search procedure (and standard search pattern) is to look at a typical search. This section supposes a fire in a large kitchen and dining area on the first floor of a two-story dwelling. Standard procedures and the search pattern described apply to all buildings.

Search begins immediately. Companies arriving at the scene immediately size up the fire situation. Engine companies obtain water and get attack lines into the house, to cut off the fire and then hit it directly. Fire fighters on the lines, by getting low, can probably see some clear area over the floor, and they check for victims near the fire. The stairway and the upper floor will be full of smoke and gases. An immediate attempt should be made to get truck crews with self-contained breathing apparatus to the upper floor. If the area is tenable, they can begin searching for victims. If the area is untenable because of the intense heat, ventilation must begin from the outside.

Windows on the lee side should be knocked out first. One fire fighter with a short ladder can quickly knock out enough second-floor windows to make a big difference (Figure 3.6).

Search pattern. As soon as the second floor is tenable, the search begins. According to standard procedure, the most dangerous area — directly over the fire — is the first place to be searched.

On reaching the top of the stairway, search personnel will have to turn in one direction or the other to get over the fire. This turn sets up the basic search pattern. As the search proceeds, these crews *keep turning in the*

Figure 3.6. In a search of a typical two-story dwelling with the fire on the first floor, search of the upper floor must begin immediately.

original direction as they go in and out of rooms. For instance, suppose that a fire fighter engaged in the search turns left at the top of the stairs to move down the hallway and get over the fire. Then, according to the pattern, the next turn is left again to move into a room. On coming out of the room, the fire fighter again turns left and moves to the next room. This left-turning path is continued around the hall and back to the stairs (Figure 3.7).

Others searching should follow the same pattern. Whenever possible, at least two fire fighters should be assigned to search in an area; one takes the first room found, and the other moves on to the next room. As they proceed with the search pattern, they should check each room that is not marked as having been searched (as discussed below). They should attempt to keep track of each other by touch, sight, verbally, and by listening for the sound of the other's breathing apparatus. Each should be alert for a call for help from the other.

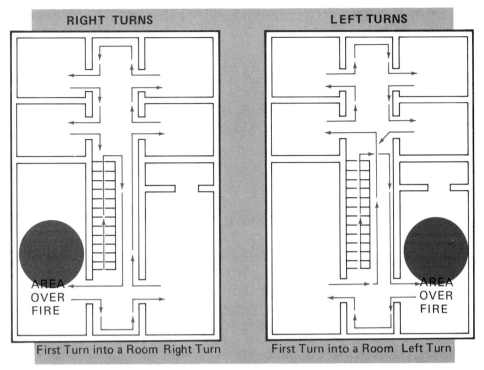

Figure 3.7. Search of the floor above the fire should follow a definite pattern.

Figure 3.8. Closets, bathrooms, and spaces under and between beds must be checked for possible victims. The tools truck crews carry can be aids in probing for victims.

Figure 3.9. When finding that fire has entered a room, the room should be searched if possible and the door shut to isolate the fire.

Areas to be searched. Corridors and halls should be checked thoroughly, as should the open area of each room. In addition, bathrooms, closets, and the spaces behind large chairs and under beds should be checked. People — especially children — often seek protection in these places. Areas near windows should be checked for victims overcome while attempting to reach a window. Everyone engaged in the search should carry a tool such as an axe, halligan or claw tool, with which these areas can be probed and which can be used for venting and forcing locked doors. Each person making a search should also carry a handlight (Figure 3.8).

If the fire has extended into a room, that room should also be searched if at all possible. Then the door should be shut to isolate the fire (Figure 3.9). Standard procedures for engine companies should include a line over the fire for just this situation. With a stream available, there is a better chance for the search to proceed quickly and safely.

If a room is not involved with fire but contains heat and smoke, it should be vented. Its door should be opened, and its windows opened or knocked out, to clear the area for a more effective search. Windows in hallways and corridors should be opened or removed.

Indicating that a room has been searched. Establish a standard way of indicating that a room has been searched, so there is no duplication of effort — at least not in the initial search operation. When the door to a room is to be left open to vent through the room, an effective indication is to place a piece of light furniture in the doorway. A chair, footstool, end table, lamp table or anything else that can be quickly dragged or carried into position will do. The piece should be set on its side so that the legs are pointing out of the room (Figure 3.10), since there is little chance that a piece of furniture could be knocked into such a position accidentally. This is quickly done and easily recognized by other search personnel.

If fire has extended into a room, or if there is danger of this happening, the door must be closed after the room is searched. Then a piece of cloth should

Figure 3.10. A chair, lamp table, footstool or similar piece of furniture should be placed in the doorway with the legs out to show that the room has been searched.

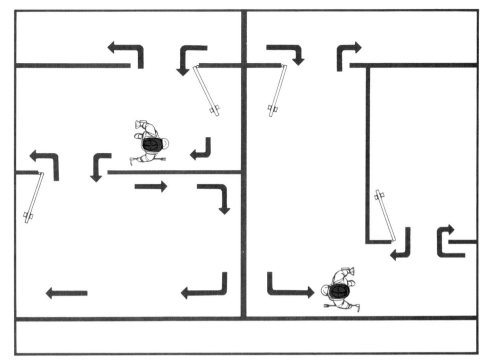

Figure 3.11. When turning into an office, apartment or similar area, fire fighters should keep one shoulder to the wall: left if left turn is made upon entering, right if initial turn is to the right.

be placed against the door jamb, at about doorknob height, and the door pulled shut against it. The corner of a bedsheet, a towel, a pillow case, a tablecloth, or an item of clothing can be used for this purpose. In this situation, a line should be called for immediately.

Some departments have had success with placing tags on the doors of inspected rooms, apartments and offices. Whichever type of search indication is used, fire fighters need not enter a room to find that it has already been searched.

Other structures. The standard search procedures and the search pattern described here apply to other buildings as well. The search pattern may be used in a single-story dwelling, an apartment house, or an office building.

Once they turn off a corridor into an apartment or office, the fire fighters engaged in the search should follow the same pattern within the unit. When they leave the unit, they should retain the pattern in the corridor and when they enter the next unit. If they enter an office unit with a right turn, they place their right side to the wall and must then keep their right side to the wall as they work their way through the unit and back to the door leading to the hall. If they enter with a left turn, they should keep their left side to the wall as they search the unit and work their way back to the door. When they leave the unit, they must retain this pattern in the hall and in entering the next office unit (Figure 3.11).

It is important that search personnel leave an office or work area through the same doorway used to enter it. Otherwise, a part of the unit containing trapped victims could be overlooked. If fire conditions force search teams to leave by a different door, they should report this fact immediately. Hoselines should be quickly advanced to the area and the unit searched thoroughly (Figure 3.12).

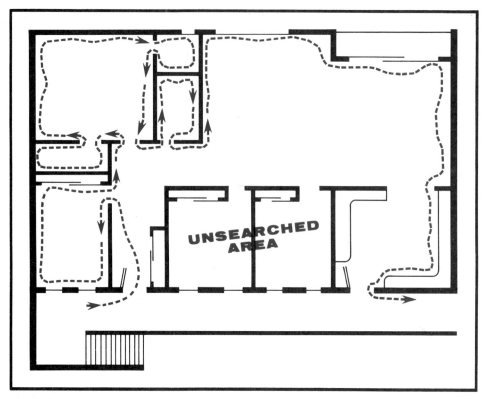

Figure 3.12. During search operations, the same door should be used to enter and exit a unit in order to prevent missing some of the rooms.

The larger the structure, the more personnel must be assigned to search and rescue. First alarm assignments must be adjusted to ensure adequate response in each area of a department's territory. Search and rescue are truck company functions; engine companies should be free to direct all their efforts to advancing hoselines in support of rescue operations. However, where truck companies do not respond (or do not have sufficient personnel), engine company personnel must carry out search and rescue operations. With a standard procedure and proper training, assignments can be adjusted to cover any situation.

Search Techniques

Search personnel must work as quickly as possible, but they must also be careful to avoid unnecessary injury to themselves and victims. This is especially true when searching near or above the fire, because although the fire may have extended into such areas, it may be hidden by doors to rooms or to sections of the building.

Doors. Before any door is opened, it should be checked to see if it or its knob is hot. A very hot knob usually indicates that there is fire on the other side of the door. Some heat in either the knob or the door probably means that the room beyond the door is full of smoke and gases, but not fire. The gases could ignite suddenly (and explosively) if the door is opened quickly. In such situations, the safest course is to first determine which way the door opens. Most doors between hallways and apartments or offices open into the unit — that is, away from the entering fire fighter. Inside the unit, doors may open in either direction. If the hinges are on the outside, then the door opens out toward the fire fighter. Some modern doors with flush, hidden hinges give no indication of which way they open. However, if the door jamb does not cover the edge of the door, then the door opens out. If the edge of the door is covered, it opens into the unit.

Doors that open out are the most dangerous to fire fighters, but they must be opened during the search. If an outward-opening door shows some heat, the fire fighter should get low and place his full body weight against the door. Then he should release the lock slowly and allow the door to open slightly. As the lock is released, the fire fighter may feel a strong push against the door from the inside. This indicates that the room is full of smoke and gas, possibly along with fire. The fire fighter should close the door immediately and call for an attack line, and the door should not be opened again until the line is in position. The same action should be taken when there is no push against the door, but the room is seen to be full of fire.

If the door opens into the room, the lock should be released slowly and the door eased in. Again, if there is a strong push against the door, or if the room is heavily involved with fire, the door should be shut and an attack line brought into position. Otherwise, the unit or room can be entered and searched.

Victims. Occupants of a fire building will try to escape through doors, windows, fire escapes, halls, and stairways leading out of the building. Fire fighters should look for overcome victims near and in such places. In particular, fire fighters must be sure that the push against an inward-opening door is due to smoke and gas, and not a victim lying against the door.

The pressure of smoke and gases is distributed evenly over the entire door. It is felt from the moment the door is first opened. If a victim is lying against the door, the pressure will be felt more at the bottom than the top. If the victim

Figure 3.13. When a door will open only part way, do not attempt to force it open further. Drop down to see if a victim is blocking the door. If so, get the victim out of the room and the building if possible. Search for others.

is near the door, it will open easily at first, and the fire fighter will feel a bump as the door hits the victim.

A fire fighter who finds both conditions — indicating that a victim is lying near the door of a room filled with smoke and gas — should attempt to open the door enough to remove the victim and then shut the door. Just getting the victim out of a smoke-filled room and into a less-charged hallway might mean the difference between life and death. If necessary, the fire fighter should immediately call for help, an attack line, or whatever is needed to get the victim out of the unit. The unit should then be searched for other victims.

When a door opens inward easily at first and then bumps an object, the fire fighter must assume that there is a victim just inside the door (few people place their furniture so that it blocks doorways). The door must be opened and the victim removed, but the fire fighter should not beat on the door with tools, body, or boots since this could only add to injuries as the door is slammed into the victim. Instead, the fire fighter should drop down and feel around the inside or probe with tools to find the victim. If possible, get the victim away from the door, open it, and remove the victim from the room (Figure 3.13).

Great physical effort is often required to move a victim away from a door. The fire fighter must work from an awkward position with a door in the way and must move a dead weight. If help is needed, call for it quickly.

If enough fire fighters are available, victims should be removed from the building while the search continues; otherwise, victims should be removed and then the search resumed. In larger structures, the latter course of action can cause too much delay; in such cases, it might be better to get the victim into the hallway, vent the hallway, and — if possible — vent the room or unit and continue the search. This is especially effective if the victim is found near or at the door and is still breathing.

If the victim is found deep within an apartment or a large room or office, the best course of action might be to punch through a partition into an adjoining apartment or room. The victim can then be moved into a less-charged area and the search resumed (Figure 3.14).

Visibility. When smoke reduces visibility during search operations, fire fighters should stay low and move quickly on hands and knees. The hands and lower legs should be used to feel for victims and for obstructions or holes in the flooring.

In hallways and corridors, the walls can serve as directional guides. As search personnel feel along hallway walls, the locations of doors will become obvious. Inside the rooms of apartments and office suites, the hands-and-knees position will keep fire fighters from tripping over furniture and other obstacles and will help in locating victims.

Windows should be opened or removed as they are encountered. This helps relieve the visibility problem and makes it easier for search personnel to find and reach victims. It also reduces the danger (from heat, smoke and gases) to fire fighters and victims alike.

Ventilation techniques and operations are discussed in detail in the next two chapters.

SUMMARY

Rescue is the prime objective of fire companies and the basic reason for their existence. Fire fighters must expect some personal risk during rescue

Figure 3.14. In extreme situations, or when deep in an apartment, it may be advantageous to force walls or partitions to get victims into a less-charged area or to a window.

operations. To minimize this danger, the operations required in rescue situations should be made part of standard operating procedures, and fire fighters should receive continual training in these operations. These procedures should include a standard search pattern that is simple to perform and thorough in its coverage of the fire building.

Search and ventilation are the main functions of truck companies in a rescue situation, along with forcible entry and laddering. However, because any fire fighter might be required to conduct a search, they all should know the standard search procedures.

VENTILATION TECHNIQUES 4

Ventilation is the process of making openings in a fire building or exposure to allow heat and the products of combustion to leave the building. Because of the design of their apparatus, and the tools and equipment it carries, truck companies are usually assigned the job of ventilation. It is one of their most important duties.

Ventilation contributes directly to the accomplishment of the basic fire fighting objectives by

- Reducing the danger to trapped occupants and thus extending the time available to fire fighters for rescue operations
- Increasing visibility, both for fire fighters and occupants, thereby decreasing the danger inherent in other fireground operations and increasing fireground efficiency
- Permitting quicker and easier entry to allow search operations or to advance lines
- Minimizing the time required to locate the seat of the fire
- Minimizing the time required by truck crews to find areas to which fire has spread within the building
- Decreasing or stopping the spread of fire
- Reducing the chance of flashover or backdraft

Which of these results is produced will depend on such factors as the size and type of occupancy involved, the extent and location of the fire, and whether the fire is free-burning or smoldering. However, when properly performed, ventilation will increase the effectiveness of most fireground operations (Figure 4.1).

In spite of the benefits of ventilation, many fire departments hesitate to ventilate buildings involved with working fires. They note that ventilation techniques require *doing* damage to a building, whereas one goal of fire departments is to *limit* damage. The fact of the matter is that the small

Figure 4.1. Ventilation, if carried out properly, will increase the effectiveness of fireground operations.

amount of damage done during ventilation operations ultimately results in a much larger *reduction* in fire damage. Even more important, ventilation directly aids in attaining the primary objective of fire fighting operations: saving lives.

The benefits of ventilation far outweigh its disadvantages. The principle is simple and straightforward, and the techniques are easy to learn and use. This chapter deals with the methods by which fire buildings can be opened for ventilation. Chapter 5 discusses the use of these methods in ventilation operations.

BASIC PRINCIPLES

In Chapter 1 it was explained that fire travel by convection presents the greatest fire fighting problem. It is by convection that hot air, smoke, heated gases and burning embers move through a building, traveling vertically where possible and spreading horizontally when vertical pathways are blocked. If the horizontal pathways are also blocked, combustion products build up and begin to fill the building from the top down. They limit visibility within the fire building, cause death by asphyxiation, and ignite secondary fires that can be more severe than the original fire.

Figure 4.2 shows a multistory building with a working fire on a lower floor. The fire floor is filled with hot smoke and gases, but these combustion products have also been convected up from the fire floor through natural channels in the building. They have accumulated under the roof, which blocks their vertical pathways, and have then moved horizontally. The appearance of such a building has led to the use of the term "mushrooming" to describe this condition.

As the fire burns, hot air and combustion products will continue to rise to the roof. More and more heat will be pumped up, and eventually something will ignite. There will then be two fires in the building, separated by several uninvolved stories.

To prevent the secondary fire from igniting, fire fighters must get rid of the accumulation under the roof by making an opening through which the hot air

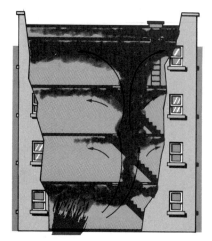

Figure 4.2. In a multistory building, combustion products convected up from a fire on a lower floor create a condition known as "mushrooming."

and combustion products can escape. To make use of the natural tendency of heated products to rise, this opening should be above the accumulation. In a one-story building, or if the fire is on the top floor of a multistory building, at least one ventilation opening should be made at the top of the building directly above the fire where the situation is most acute. In other situations, openings should be made at the tops of the vertical shafts through which combustion products are rising.

At the same time, or immediately afterward, the accumulation on the fire floor should be relieved to prevent combustion products from being pushed up through the building. Openings should be made at the fire floor, usually through windows and doors, and should lead directly from the accumulations of heat and combustion products to the outside of the building. Similar action should be taken on the floor above the fire, as soon as possible, so the area can be searched and lines advanced to check fire extension.

This, then, is the general rule behind all ventilation operations: open the fire building in such a way that all accumulations of heat and combustion products will leave the building by natural convection. The principle is obvious; the results, in terms of the reduction of deaths, injuries, property damage and fire spread, are immeasurable.

NATURAL OPENINGS

When opening the roof is the only way to get rid of accumulated combustion products (or to prevent them from collecting), there should be no hesitation in making the opening, but this is not always necessary. In many cases, a building can be effectively ventilated through natural openings — "built-in" construction features that can be quickly opened and easily repaired. Windows are natural openings, as are skylights, roof hatches, ventilators and penthouses. The effectiveness of using natural openings for ventilation depends on their location in relation to the fire and on the pathways open to the combustion products.

Not all of the natural openings discussed in the following pages will be found on every building. Truck companies should, through prefire inspection, determine which are available and how they can best be used for ventilation. Truck crews should be able to recognize natural roof openings and the building areas served by these openings, and should know the most efficient methods for uncovering them with standard hand tools.

Windows

Along with the location of the fire, the construction, type and size of the fire building determine how it should be vented. What may be required to properly ventilate one building might be unnecessary in another. For example, unless the fire is in the attic, roof ventilation is rarely necessary in one- or two-family dwellings since these structures can usually be vented through windows.

When time permits, the windows should simply be opened. Double-hung windows should be opened about two-thirds down from the top and one-third up from the bottom. Other types of windows should be opened as much as possible. If a window is equipped with a storm window, it, too, must be opened or knocked out.

Shades, venetian blinds, drapes, curtains and other window coverings must be moved away from the window. If they cannot be raised or moved to the side quickly, they should be pulled down.

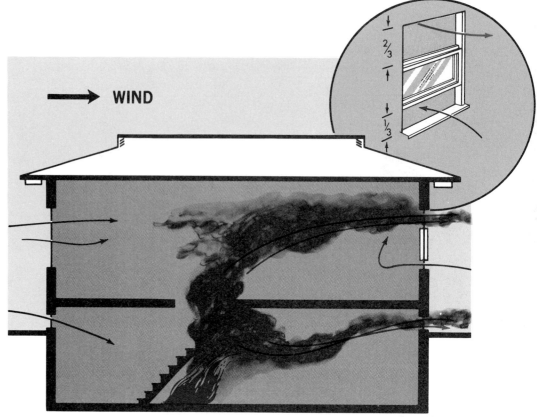

Figure 4.3. When using windows to ventilate a building, those on the leeward side should be opened first.

If there is not enough time to open windows and storm windows, they should be knocked out with a truck tool. Use as little force as possible, so no broken glass will endanger fire fighters who may be working outside. Fire fighters outside the fire structure must be wary of the possibility of flying glass.

In some cases, truck crews cannot get inside the building to open windows. Then the windows should be knocked out from the outside with ground ladders, or from ground ladders or aerial units. These operations are covered in Chapters 8 and 9.

Effects of wind. When wind is a factor, the windows on the leeward side of the building should be opened first (Figure 4.3). Then the windows on the windward side can be opened to allow the wind to blow combustion products out of the building. If the windward side is opened first, the wind will churn smoke and gas around the interior until the leeward side is opened, but opening the windows in proper sequence will create effective cross ventilation.

Window and roof ventilation. When the roof or some roof features must be opened for venting, the windows on the top floor should be opened or knocked out *after the roof is opened*. If this cannot be done from inside, the windows can be knocked out with tools from the roof or from a ladder, fire escape, porch or balcony (Figure 4.4).

When the windows on several stories must be opened or knocked out, fire fighters should begin at the top and work down. Fire fighters going up to the top floor on a fire escape, for example, should not open windows as they go up

Figure 4.4. Following roof ventilation, the top floor windows should be opened.

since this would allow smoke to flow out of the windows and cover both them and the windows above. Flames might also issue from the windows and trap the fire fighters.

If the fire or trapped gases should blow out a window above crews who are working their way up, they should seek shelter on the lower floors or in the nearest window area to avoid flying glass and other debris. They should never look up to see what is happening. Instead, they should take cover until the danger is past.

Natural Roof Openings

In multistory buildings, vertical shafts carry stairways, elevators, dumbwaiters, electric wiring, heating ducts, and plumbing and sewer pipes. These shafts extend through the full height of the building; convected heat, smoke and gases will rise within and around them when the building is involved in fire. If this happens and the shafts are not opened at the roof, fire will travel horizontally from the tops of the shafts, ignite the top of the structure, and then work down.

The slight pressure that will develop at the top of the shaft will force heat, smoke and gases throughout the upper parts of the building, pushing combustion products into other shafts, into hallways, and into apartments and offices through doors and transoms. The greater the buildup of combustion products, the greater is the probability of fire on the upper floors.

The shafts are capped, at the roof, with various types of closures; these can be removed to make effective openings into the building.

Skylights

The positioning of skylights can give fire fighters an idea of the layout of the building under them. For instance, in an apartment or office building, a row of skylights from front to rear is most likely located over the top-floor corridor. In shops or factories, a line of skylights is usually placed over a work area. In apartments, office buildings and similar structures, individual skylights are often located over stairways, corridors, elevator shafts, air shafts and bathrooms. Those placed over bathrooms usually have louvered ends to allow normal venting. Skylights located over other features are usually solid.

The area immediately below a skylight is usually boxed in so the cockloft or attic space is effectively separated from the skylight. Thus, when the skylight

Figure 4.5. Skylights can be cut or pried loose and removed, or turned over using one side of the flashing as a hinge. The glass can either be removed or knocked out.

is opened, the building proper will be ventilated, but the space just below the roof will not. To ventilate this space, the roof or the boxed-in area must be opened.

Very often, fire fighters will be working in the area below a skylight. Therefore, before the skylight is removed or its glass knocked out, some warning should be given to those below by banging on the roof at the base of the skylight or on the sides of the skylight at the roof line. Fire fighters working below the skylight must be aware of the meaning of this signal and should take cover. The signal should be given even if truck crews intend only to remove or tip the skylight, since the glass might accidentally break in the process.

Opening skylights. There are three ways to open a skylight for venting. The preferred method is to either lift the skylight from its opening or tip it over onto the roof. The flashing that joins the skylight to the roof must first be cut or pried away. If this can be done quickly on all four sides, the skylight can be lifted off the opening (Figure 4.5A). Otherwise, the flashing should be removed from the two short sides and from one long side and the skylight then tipped over using the fourth side as a pivot (Figure 4.5B).

Some skylights are not mounted directly on the roof, but are instead placed on a wooden foundation 6 inches or so above the roof. The flashing is then attached to the wooden foundation rather than to the roof. In this case, too, the flashing must be cut or pried loose before the skylight can be removed.

To ensure that fire fighters do not fall through the opening where a skylight has been lifted away, the skylight should be laid on the roof *upside down* (Figure 4.5A) to serve as a warning. Then it will not be mistaken in smoke or darkness as being in place over its opening.

If for some reason the skylight cannot be lifted or tipped, the glass can be removed. The least damage is done by peeling back the metal stripping along the bottom edge of the glass and sliding the panes out (Figure 4.5C). If the glass will not come out easily and quickly, it must be knocked out. Although this method is the least desirable, it might be the only way to quickly open the

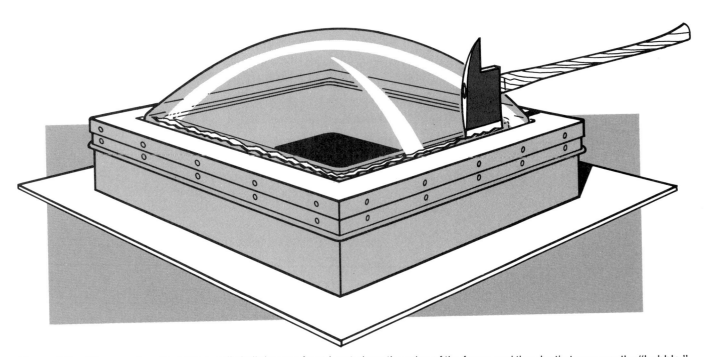

Figure 4.6. When modern "bubble-type" skylights are found, cut along the edge of the frame and the plastic to remove the "bubble."

skylight for venting. The skylights installed over garages, stores, shops, factories and other large open areas can be huge and very heavy. One or two fire fighters will not be able to lift or tip this type of skylight and will have to remove or knock out the glass.

Plastic skylights and roof panels. Various sizes of clear plastic "bubble-type" skylights are installed in some modern structures. These are made by setting a one-piece unbreakable plastic bubble in a frame. Plastic skylights are mounted in the same way as glass skylights and can be lifted or tipped in much the same manner. When a plastic skylight cannot be removed quickly, the frame should be cut where it meets the plastic, and the bubble pried up (Figure 4.6).

Clear, frosted or colored plastic panels are sometimes placed in a roof to serve as a simple skylight. Since these panels are usually weak and will not support much weight, truck crews should avoid stepping on them. These panels are used mainly in the gable roofs of modern noncombustible buildings, such as warehouses, factories, garages and shops. The roof itself is often constructed of corrugated metal with no ceiling below. The plastic panels can be pulled up after the roof is cut, or pried up along one edge of the panels (Figure 4.7).

Effects of wind. Any large opening in the roof, such as a skylight opening, will allow fire, heat and smoke to rise up through the roof. For this reason, truck crews opening a roof or any roof feature should keep their backs or sides to the wind. If they face the wind, their faces may be subjected to a blast of fire, heat, or smoke and gases.

Openings below the skylight. Once the skylight is removed or knocked out, smoke should flow freely from the opening. If it does not, there might be a swinging transom or a glass panel at the ceiling line below the skylight. The transom should be opened if possible or its glass knocked out. A glass panel should be removed, if this can be done quickly; otherwise, it should be knocked out. Do this carefully, to avoid the heat and smoke that will flow out of the opening.

If the area below the opened skylight is boxed in, the cockloft must be vented through an opening in the roof. A roof scuttle (see below) can be used for this purpose, if it is located in the right spot and is not boxed in like the

Figure 4.7. Plastic panels are sometimes placed in the roof to serve as skylights.

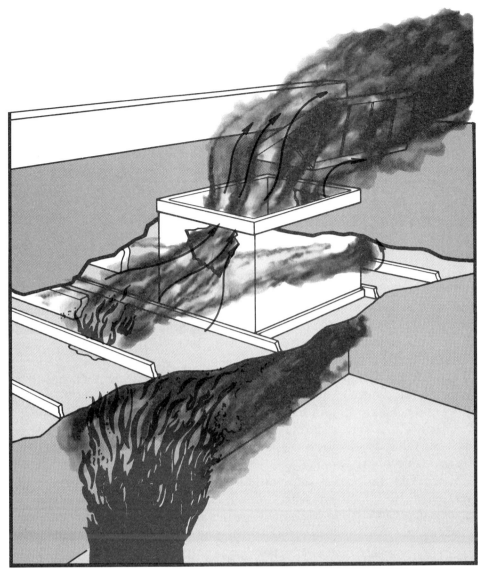

Figure 4.8. To vent the cockloft or attic, the roof or the closed-in area beneath the skylight must be opened. Removing the skylight alone will not vent the area under the roof in such cases.

skylight. If no such roof feature is available, it is necessary to open the boxed-in area below the skylight or to cut the roof open around the skylight (Figure 4.8).

If the fire is directly under the roof (that is, on the top floor of the building), a roof opening should be made as close over the fire as safety allows. Otherwise, the opening will draw the fire across the top of the fire floor, under the roof or ceiling.

When the roof is separated from the fire by at least one story, the roof can be opened around the skylight after the skylight is opened. In this case, the boxed-in area can be opened instead of the roof, but if smoke and heat issuing from the skylight shaft make this impossible, then the roof will have to be opened.

Roof Scuttles (Hatches)

A scuttle is placed in a roof to allow access to the roof from inside the building. Sometimes a ladder is built into the wall below the scuttle. In multiple occupancies, scuttles are usually located above and at one end of the top-floor corridors. In stores and other business establishments, they are usually located at the rear of the building (Figure 4.9).

Scuttles vary in size but are usually square in shape. A scuttle consists of a wooden cover that fits tightly over a raised wooden support attached to the roof. The cover is encased in roofing metal and is sometimes tarred. It is often held in place only by the support. Over the years, rust, corrosion, grit and dirt which get between the cover and support can tighten the seal. To prevent burglaries, the covers are sometimes bolted or padlocked in place on the inside; they might also be secured by steel bars.

Opening scuttles. A cover that is not securely locked in place can be pried off with a pickaxe, halligan tool, claw tool or similar device. If a cover cannot be removed quickly, the top of the cover should be cut out. It can easily be replaced.

Opening below the scuttle. As with skylights, the area below a scuttle might be boxed in to separate it from the cockloft. Once the cover is removed or cut, truck crews should check to see if the scuttle is closed at ceiling level. If smoke pours out, these areas will have to be probed with tools.

If the scuttle is open at the ceiling level but boxed in through the cockloft, the building is being vented. If smoke is coming out but there is a panel at the ceiling level, the cockloft is being vented. If there is no closure at the ceiling level or the cockloft, both the building and the cockloft are being vented.

If there is a closure at the ceiling level, it must be removed or knocked out. This should be done with an axe, a ceiling hook or some other tool, depending on the depth of the opening.

If the enclosure below the scuttle is boxed in, the cockloft is not being ventilated. Such enclosures are usually made of very light tin, thin wood, or plaster that can be removed quickly; some, however, are made of a tougher material that is difficult to open.

Ventilators

There are various types, sizes and shapes of ventilators and vent pipes, and each usually serves a different purpose. The vent pipe is customarily capped by a ventilator.

Some ventilators open only into the cockloft. Others open into the top floor of the building; these are used to exchange air and equalize pressures within the building. Others may be on vertical shafts that extend the full height of the building; these usually connect into heat-producing areas, such as kitchens and laundries, and might contain fans to increase the flow of air. Still others are used to ventilate bathrooms in multiple occupancies, including apartments, office buildings, hotels, and motels; these usually contain blowers or fans. The huge ventilators on the roofs of theaters and auditoriums usually open into the main assembly area at ceiling level; their ceiling inlets are normally covered with decorative interior grilles.

Each of these ventilators can be used to vent a building. Truck company personnel should be familiar with the different types of ventilators and the type of venting each provides (Figure 4.10).

A louvered ventilator that can be rotated by the wind or by rising heat (Figure 4.11) is usually connected to a shaft that runs the full height of a multistory building. In a single-story building, this type of ventilator is usually

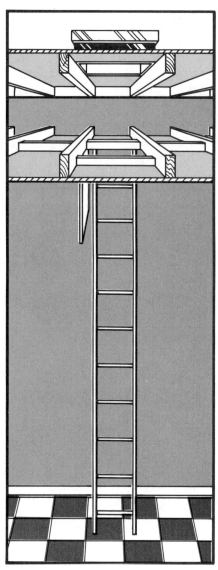

Figure 4.9. Opening scuttles can be an aid to venting. If the scuttle is locked securely, cut out the top. Once open, check for an enclosure through the attic space and for an opening at ceiling level.

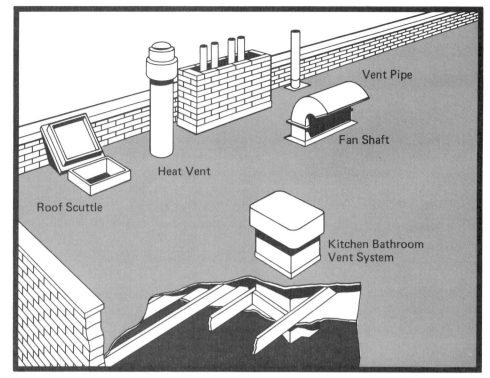

Figure 4.10. Ventilators can aid truck crews in venting a building. Fire fighters should recognize the different types.

placed directly over a heat-producing area. In a residence, it is used to vent the attic space to assist in cooling the structure.

In an apartment house, ventilators that cap the fan shafts leading from kitchens or bathrooms are normally lower than other ventilators, and they have a different appearance (Figure 4.12). They often curve as they come out of the roof, so that they point down toward the roof. Some are shaped like small skylights; they are made of roofing tin with a solid top. Their open ends are covered with hardware cloth (rat wire) to permit the venting action.

Other ventilators, the stationary type, are not connected to heat-producing areas. These consist of a pipe extending above the roof, a cover located an inch or two above the pipe, and a wide metal band that passes around the top of the pipe and the cover. The cover and band keep rain, snow and dirt from being blown into the pipe.

Removing ventilators. Smoke coming from a ventilator indicates that the fire has reached the area it serves. The ventilator should be opened to remove the restriction at the top (the weather cover).

A stationary ventilator can be removed by getting a hold on the underside of the metal band with the pick of an axe, a halligan tool or a claw-tool hook, and pulling off the band. The cover usually comes off with the band; if not, it can be pulled away from the pipe. The amount of smoke leaving the ventilator will increase dramatically after the cover is removed (Figure 4.13).

Rotating ventilators are usually stronger than the stationary type. If a rotating ventilator is difficult to remove, it should be broken off at the roof line. Although this is not a desirable action, since it leaves a hole in the roof, it may be necessary.

The low ventilators over fan shafts can usually be pulled up from the shafts. The curved type should be pulled off at the roof line (Figure 4.14).

Figure 4.11.

Figure 4.12.

 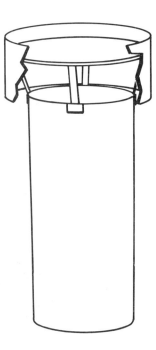

Figure 4.13. Smoke issuing from a ventilator is an indication that it should be opened for full effect. Use a tool to remove the top.

Sheet-metal shafts. It is especially important that truck crews be able to recognize ventilators that cover sheet-metal shafts. Since the metal is usually very thin, the heat of a fire traveling through such a shaft can ignite combustible materials in contact with the shaft. In addition, the fire might extend through seams in the shaft and ignite building features around it. This is especially true of grease duct vents in restaurant and hotel kitchens.

The roof area around this type of shaft should be checked for heat if the building is involved with fire. If the roof is hot and the tar shiny, or if the tar is dry on a wet day, the roof should be opened around the shaft or its ventilator. This will allow fire around the shaft to be drawn up and out of the building and will deter lateral fire spread.

Plumbing system vent pipes. Each separate plumbing system in a building has a vent pipe that extends through the roof. These pipes are about 2 inches in diameter, so they can be run vertically through walls or partitions. Since they pass through each floor of a multistory building, they must be considered as vertical shafts and as possible channels for the spread of fire and combustion products.

The roof area around each plumbing system vent pipe should be checked. If it shows signs of heat or smoke, the roof should be opened around the pipe.

Penthouses (stairway covers; bulkhead doors). A penthouse is a small hutlike enclosure built over a stairwell that allows the stairs to extend up to the roof level and is tall enough to permit a person to walk (rather than climb) onto the roof. The penthouse has a full-sized door at the roof level, and there can also be a door at the top floor. These doors can be kept locked or unlocked. The penthouse itself might have windows in its sides, and a skylight on its roof. It is mostly found in older apartment and office buildings (Figure 4.15).

Figure 4.14. Ventilators over fan shafts should be removed to assist in venting the building. Fans should be shut down.

Figure 4.15. Opening the penthouse can aid in ventilation of stairway and corridors.

Since a penthouse covers an open stairwell, it can be used to ventilate the entire height of the stairwell. Opening the penthouse will also ventilate corridors and hallways that are open to the stairs. Because the stairs and corridors will be used by occupants attempting to leave the fire building, and by engine crews advancing their attack lines, penthouses should be opened as soon as possible.

Heavy smoke showing around a penthouse usually indicates that it is open to the stairway. The roof door is opened in the same way as any other door — by force, if necessary. If there is a closed door at the top floor level, it must be opened immediately.

Before the lower door is opened, it should be checked for signs of heat. If the door is very hot, a blast of heat might hit the fire fighters when they open it, and smoke might obscure their vision. They may have difficulty getting back up the stairs to the roof. Instead of opening a hot door, the fire fighters should try to force a hole through the top of the door to check the stairwell. If it is filled with smoke or heavily involved, they should open the hole to about one-third of the door height and *get out*. This will provide venting for the stairwell, but not endanger the fire fighters as much as opening the entire door.

If the door is of heavy construction and difficult to break through, it should be left closed. The fire fighters should return to the roof, and make an opening in the roof over the stairwell. This will ventilate the stairwell just in front of the lower door.

Machinery Covers

Small, boxlike enclosures are sometimes found on the roofs of office buildings, apartments and commercial buildings. These machinery covers usually house dumbwaiter pulleys or motor assemblies, the valves for heating and air conditioning systems, and similar items. They are usually made of wood and covered with roofing metal; sometimes they are tarred. They can also be equipped with ventilators.

When fire fighters detect smoke issuing from a machinery cover, or heat in its walls, they should vent it. This can usually be done simply by opening its service door.

Elevator Houses

If there is an elevator in a building, there is probably an elevator house on the roof. The elevator house contains the motors and electric switches that control the elevators. It is located directly above the elevator shaft and is open to the shaft.

By ventilating the elevator house, fire fighters can reduce the accumulation of heat and smoke in the shaft, making the elevators safer for use. At the same time, removing excessive heat from around the motors and switches in the house lessens the chance of machine damage, which could result in erratic operation of the elevators. Even if there is fire in the elevator shaft, and the elevators are not being used, the elevator house should be ventilated to help draw the fire out of the building and deter lateral fire spread.

Elevator houses vary in size, depending on the number of elevators in the building. They range from small structures no bigger than an elevator car to huge affairs serving several elevators. The larger houses often have windows that can be opened or knocked out to vent the house and the shaft. The door to the elevator house also can be opened (by force, if necessary) to provide ventilation. If there is a skylight on the roof of an elevator house, it, too, can be opened for ventilation. However, these skylights are often difficult to reach without a short ladder (which must be brought to the roof for that purpose). It is usually quicker to vent through the door.

Air Shafts

Air shafts are usually found in older buildings. They are most often rectangular in cross section and can have three or four sides. Shafts with only three sides normally are open toward the rear of the building.

Air shafts might be located entirely within one building or between two row structures. Their purpose is to allow light and air to enter the inner rooms. Windows in these rooms open into the air shaft at each floor. Most air shafts are not covered at the roof, so the inner rooms of the building can be ventilated by opening these windows (Figure 4.16).

Some four-sided shafts have skylight-type coverings at the roof. These are usually referred to as "light shafts." If fire has reached such a shaft, the skylight should be removed or opened so heat will not build up in the shaft and be forced into the rooms around it. The fire should be reported immediately, so hoselines can be placed to cut off its spread. The skylight should also be opened if the shaft is being used to vent inner rooms.

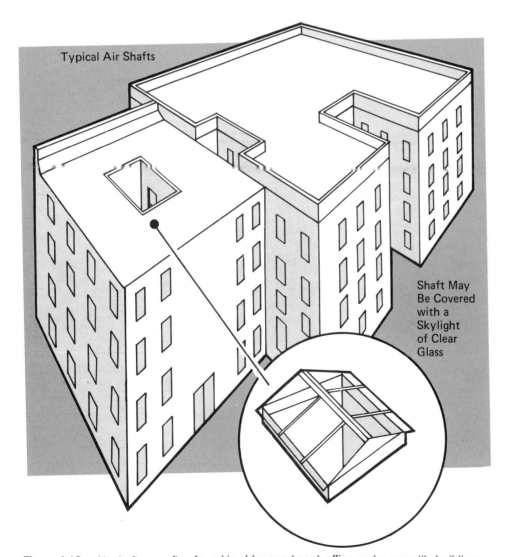

Figure 4.16. Air shafts are often found in older apartment, office and mercantile buildings.

If the skylight cannot be lifted or tipped quickly, the glass panes should be removed or knocked out. The shaft will not be used by fire fighters or escaping occupants, so there is little danger of glass falling on anyone.

Prefire Inspection

Although few, if any, buildings will contain all the natural openings discussed in this section, every building will have some of them. The only sure way to know which natural openings can be used in fighting a fire is by making prefire inspections.

A check through one apartment building in a housing project will usually enable truck crews to determine which roof features vent the kitchen fan shafts and how best to handle these roof features. A visual check for vertical story-to-story rows of small frosted windows will indicate the locations of bathrooms and the accompanying pipe shafts that rise up through the building.

Prefire inspection allows truck crews to associate a particular row of kitchens with the roof feature that vents them and to match rows of bathrooms with their roof vent pipes and fan-shaft ventilators. The information gained in one building will apply to all similar buildings in the project.

Prefire inspection can save time and effort on the fireground; it is as important to ventilation as it is to any other fire fighting operation.

Figure 4.17. If it is necessary to open a roof, a large hole should be made. Care should be taken to avoid cutting structural members.

CUTTING THROUGH ROOFS

At times, the only way to properly ventilate part or all of a building is to cut a hole in the roof. A roof made of boards under standard roofing materials can be cut with axes, and the cut areas forced with halligan or claw tools or the pick of an axe. A plywood roof should be cut with a power saw, if available, since it is difficult and time consuming to cut through plywood by hand. In either case, fire fighters should be careful not to cut joists or other structural members; this will weaken the area in which they are working (Figure 4.17).

A single large hole is more effective than several small holes and is also safer for fire fighters operating on the roof. One 4 X 8-foot hole has twice the area of four 2 X 2-foot holes and requires less cutting.

All the roof boards should be cut through before any of them are pulled up. If a board is pulled up as soon as it is cut, the smoke pouring through the small hole could drive fire fighters from their position, making it impossible to complete ventilation. When pulling up the cut boards, truck crews should keep their backs to the wind.

When all the boards have been ripped up, the ceiling below should be knocked down with a pike pole or similar tool. The ceiling hole should be made as large as the roof hole. There is little sense in making a roof hole of the proper size and then restricting ventilation with a tiny ceiling hole.

Special care must be taken when the fire is immediately below the roof — that is, in the attic under a gabled roof, or in the cockloft under a flat roof. In such cases, the roof must be opened as close as possible to the seat of the fire. Otherwise, the fire, along with heat and combustion products, will be drawn across the top of the building to the opening.

Heat rising from the fire will develop a *hot spot* on the roof which will indicate the location of the greatest concentration of fire (Figure 4.18).

If the roof is flat, the opening should be made at the hot spot or as close to it as fire fighters can get safely. If the roof is gabled and the hot spot is at the peak, the opening should again be made at the hot spot. Otherwise, a gabled roof should be opened so that the hole extends from the hot spot toward the peak of the roof to speed up the venting of accumulated smoke and gases

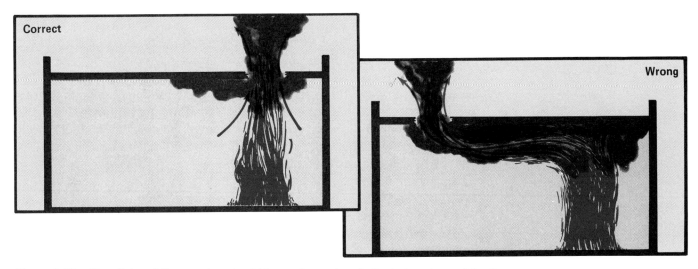

Figure 4.18. On a flat roof, the opening should be made as close to the hot spot as safety allows.

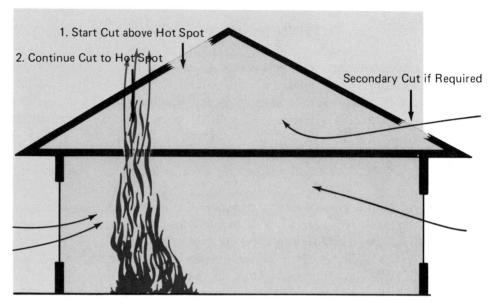

Figure 4.19. In venting a gabled roof, start cutting near the peak down to the hot spot.

(Figure 4.19). The draft can be increased by making an additional opening just above the eave line on the side of the roof opposite the original opening.

The same sort of care must be exercised when a natural roof opening is used to vent an attic or top-floor fire. By removing a scuttle or skylight that is some distance from the fire, truck crews might actually increase fire spread. If the hot spot is not near enough to a natural opening, the roof should be cut.

FORCED VENTILATION

Forced, or mechanical, ventilation is usually accomplished through the use of smoke ejectors (fans) or fog streams. Of the two, fans are the more popular because they can be used at any time and any place during fire fighting operations. A fog stream is usually used only to hasten the removal of smoke from an area after the stream has extinguished the fire there.

Smoke Ejectors

Smoke ejectors cannot be used as a substitute for the venting techniques described earlier in this chapter. The heat, smoke and gases produced by a working fire are most effectively removed by natural convection through natural or forced openings. Fans should be left on the truck when it arrives at a working fire; truck crews should perform normal venting activities and carry out other necessary duties. Truck crews carrying fans and following engine crews into the building or up stairs are contributing nothing to the fire fighting operation. They are not providing ventilation when and where it is needed, and because they cannot carry other equipment, they cannot search for victims or assist personnel in advancing hose lines. Proper ventilation is delayed, which can cause additional problems.

Fans should not be used in partially or completely confined spaces, such as attics, corridors or closed-up basements, in which there is a working fire. Under these conditions, the fans could spread the fire laterally. They can be used in confined spaces after the fire has been knocked down, but, even then, fire fighters operating hoses must keep close watch to ensure that the air movement does not fan embers into open flame.

In spite of such problems, fans can be of great help on the fireground. In some situations (discussed in the next chapter), they can constitute the major item in a venting operation. In others, they might be used to supplement more standard ventilation techniques.

Fan placement. Fans are most effective when placed where they tend to increase natural air flow. They should, therefore, be positioned in windows, doorways, roof openings, basement openings, or openings that have been made to ventilate the building.

When a fan is positioned in a window or doorway, all shades, drapes, blinds, curtains and screens should be removed to eliminate restriction of air flow. If possible, the open area *around* a fan should be closed with salvage covers or whatever materials are available; this increases the fan's efficiency by directing air to and through the opening and by preventing smoke from reentering (Figure 4.20).

Figure 4.20. For greatest effectiveness, the open area around the fan should be closed.

Fire fighters must be careful when positioning fans not to exhaust smoke into congested areas. Fans must be placed so no smoke is blown into the open windows of nearby buildings or into the intakes of heating and cooling systems.

Fans in tandem. Fans can be especially effective when used in pairs. When two fans are being used to exhaust smoke, one should be placed near an outside opening so it blows smoke out of the building; the other should be positioned inside the room so it blows smoke toward the first fan.

Two fans can also be positioned to exhaust smoke and draw in fresh air at the same time. The exhaust fan should be mounted high in a ventilation opening where smoke and gases have collected. The intake fan should be lower, at floor level if necessary, so that working fire fighters will benefit from the fresh air.

Fog Streams

Fog streams can be used to start ventilating immediately after the fire has been knocked down in a room. For this, the stream must be directed out of a window in such a way that it draws out the remaining smoke and gases.

To be most effective, the stream should be positioned so the fog pattern covers most of the window opening. This usually means that the nozzle should be held a few feet inside the window. A good way to determine where to hold the nozzle is to start by holding the open nozzle outside the window and then slowly backing it into the room while observing the movement of the smoke. At the proper nozzle position, the stream will draw the smoke out the window in large volumes.

By starting outside, the fire fighter on the line will also be able to check whether or not the water will do any damage or disrupt other operations. Fog streams should not be used for venting if they will damage items removed for protection; if they will enter adjoining buildings or damage the outside of the fire building; if they will do unnecessary damage inside the room from which they are directed; or if they will cause ice to form outside, thus creating a hazard to fire fighters.

Fog stream venting should be a short term operation. If the operation is obviously ineffective in a particular situation, other venting techniques should be used; both the personnel and the lines can probably be used to better advantage, especially if the fire force is short in numbers.

Solid streams can be used in a similar manner, either by opening the nozzle halfway or by removing the tips and using the shutoff opened fully. The broken streams created will move sufficient air to clear the area.

SUMMARY

Ventilation is of prime importance in saving lives, increasing the effectiveness of other operations, and reducing property damage.

Ventilation means opening a fire building to the outside so that heat, smoke and gases can leave the building, mainly by natural convection. In many cases, fire fighters can make use of natural openings, thereby minimizing the damage done during ventilation operations. However, when required by the fire situation, there should be no hesitation in opening the roof and top floor of a building to ventilate it. Savings in terms of deaths, injuries and property damage far outweigh the cost of repairing the building.

Among the roof features that can be opened to ventilate all or part of a building are vent pipes with ventilators, roof scuttles, stairway and machinery covers, elevator houses, and penthouses. In addition, fans and fog streams can sometimes be used to enhance ventilation.

VENTILATION OPERATIONS 5

In the previous chapter, the benefits of proper ventilation and the ways in which fire buildings can be opened to remove accumulated heat and combustion products were discussed. It is not necessary — or possible — to use every venting technique in every building. The way in which a specific fire building is vented depends on its size, construction and occupancy; the size and location of the fire; and whether the fire is free-burning or smoldering.

This chapter deals with the use of venting techniques in several types of fire situations. Not every situation is or can be included, but those that are discussed illustrate how some basic principles may be applied to any structural fire. In a particular situation, truck officers and crew members must act on the basis of their knowledge of venting techniques and their size-up of the situation.

On the fireground, ventilation duties must be performed as a part of the overall fire attack operation. When the fire is free-burning, ventilation should begin at the same time as the initial attack or as soon after it as possible. When the fire is smoldering or is suspected to be smoldering, the building must be vented *before it is entered*. Most of this chapter deals with ventilation operations for free-burning fires; the last section discusses how venting techniques are applied to smoldering fires.

ONE- AND TWO-STORY DWELLINGS

In most situations, one- and two-story dwellings can be adequately ventilated through windows. There is usually no need to vent through either natural or forced roof openings unless the fire is in the attic.

One-story Dwellings

In a single-story dwelling, truck crews should open or remove windows close to the fire, while the engine crews begin fire attack. The first windows to be opened are those through which fire or smoke is pushing out of the structure or through which fire can be seen or heard. This will immediately improve conditions in the dwelling and allow engine crews to advance their lines as necessary (Figure 5.1).

Truck crews should then enter the dwelling to search for victims and check for fire extension. At that time, they should open the other windows to complete the ventilation. The attic or cockloft should be checked for fire spread, especially in the area directly above the fire.

Two-story Dwellings

If the fire is on the first floor, the first-floor windows closest to the fire should be opened immediately. As noted above, the first windows to be opened are those that show fire or smoke. The second floor should also be vented and a search for victims and fire extension begun as soon as truck crews can enter the dwelling.

If the fire is on the second floor, that floor must be vented first. If conditions warrant, ventilation can be started by truck crews on short ladders placed outside the building. Where possible, the fire floor should also be vented from the inside.

Figure 5.1. When fire is seen behind windows, or smoke is pushing out around them and fire can be seen or heard, truck company personnel should open these windows from the outside to assist in gaining entrance.

The attic or cockloft should be entered, checked for fire, and vented if necessary. This is especially important where units of a two-family dwelling are side by side, since such dwellings often contain a single common attic across which fire can spread from one unit to the other. Since fire can also spread through the (usually flimsy) dividing wall between side-by-side units, this too should be carefully checked along its full length and height.

Attic Fires

In many small dwellings, the attic is a fairly large open area, completely or partially floored, and often loaded with stored household goods. A set of stairs usually leads to this type of attic from the rear of the house. Other dwellings have much smaller attics that are lower and only partially usable for storage. This type of attic is entered either by climbing a set of pull-down stairs, or through a scuttle at the ceiling level.

A working fire in either type of attic should be attacked from within the building, rather than through the windows. To aid engine companies, truck crews must ventilate the attic. If there is a window at each end of the attic, both windows should be opened or removed from outside. If an attic has no windows, it may have built-in louvers at each end for normal ventilation. These are usually located under the peak of the roof and can be removed to accelerate venting. If necessary for adequate venting, the roof should be opened at or near the hot spot.

Basement Fires

A working fire in the basement of a small dwelling should be ventilated through all available basement openings. In addition, the first floor should be thoroughly vented.

The venting of the basement should be coordinated with the movement of attack lines. If possible, the attack lines should be advanced through both the outside basement entrance and the first-floor basement entrance. Then, any other doors and the basement windows can be used for venting. If the fire is attacked through only a single stairway, then any other available openings can be used to vent the basement. By also venting the first floor, truck crews will aid engine personnel in positioning their lines and advancing them to the basement.

MULTIPLE-USE RESIDENTIAL AND BUSINESS BUILDINGS

These buildings involve a wide range of sizes, shapes, layouts and construction materials. Some have stores on the ground floor and apartments or offices above. Older buildings with such combinations of tenancy usually have from two to six stories and are attached; some newer shopping centers have one or two stories of offices above the ground-floor stores.

Older buildings are usually of brick-and-wood-joist construction. Their interior stairways are usually open, and elevator shafts can be open or closed. They probably contain a number of vertical shafts for dumbwaiters, utilities and ventilating systems. Newer buildings often are of fire resistant construction.

With the exception of fire resistant buildings (to be discussed shortly), these structures and buildings of similar construction, including schools, hospitals and other institutions, all require the same general ventilation procedures. When a working fire has made considerable headway in such a building, ventilation should begin at the roof.

Roof Operations

Access. In order to vent the roof, truck crews must get to it. If possible, they should use some means other than the aerial unit; this keeps the aerial available for rescue, top-floor venting and other operations. Truck personnel should not, however, attempt to use interior stairs in the fire building for access to the roof. They could find themselves in an untenable and dangerous position if the fire extends to a corridor or to the stairway itself.

If the fire building abuts a building of the same height, truck crews can climb interior stairs there to the uninvolved roof and then cross over to the roof of the fire building. They might also use fire escapes on the fire building, provided these are not crowded with occupants evacuating it (Figure 5.2). If the fire building has two or more fire escapes, truck crews might be able to reserve one of them for roof access. They must, however, make sure that the fire escape they use has a ladder to the roof.

If there is no other way to get to the roof, truck crews must use ground or aerial ladders or aerial platforms. An aerial ladder should be placed so that at

Figure 5.2. To reach the roof of the fire building, truck crews should, if possible, use means other than the aerial unit.

Figure 5.3. An aerial ladder should be placed four to five rungs above the roof edge so it can be located easily in smoke or darkness.

least four rungs extend beyond the roof (Figure 5.3). This will allow fire fighters operating in smoke or darkness to find the ladder when they need it. An aerial platform should be placed so that at least half the width of the basket extends above the roof, if the gate is on the front of the basket.

A ladder or platform used for roof access should remain in place until crews leave the roof. If it is needed for rescue, the unit must be returned to its original roof position as soon as the rescue is completed.

If visibility is poor, truck crews should probe for the roof with tools before stepping onto it (Figure 5.4). The roof is often well below the top of its surrounding wall, especially at the front of a building. A fire fighter who guesses at the position of the roof, because it cannot be seen, may be injured seriously. When they reach the roof, no matter how they got there, crews should immediately look for another way off, to be used in an emergency.

Personnel. If possible, at least two fire fighters should be sent to the roof for the venting operation. They can work together and keep track of each other. If one is injured, the other will be there to help or at least to call for help. When two fire fighters cannot be spared and the building must be ventilated, one will have to do the job alone. The officer who makes the assignment will be aware that the fire fighter is alone. The officer should be watching for the fire fighter's return; if this does not happen in a reasonable length of time,

Figure 5.4. Before getting off the aerial unit, fire fighters must determine the actual location of the roof to avoid serious injury.

others should be sent to search. These fire fighters should have at least one portable radio, to report conditions if necessary and to maintain contact for their own safety.

Venting

Roof venting should begin with whatever natural openings are available. A roof feature that shows smoke should be opened first. Skylights, scuttles and penthouses should then be quickly opened. The tops of vertical shafts should be checked; a shaft should be opened if heat or smoke can be detected around it (Figure 5.5).

Once the roof has been vented, the top floor should be opened. To do so, truck crews should use hand tools to knock out the tops of the windows from the roof or from fire escapes or porches. If conditions permit, truck crews can work their way down to the top floor from the roof, using an interior stairway or a scuttle that has a ladder. They can then open windows from the inside, at the same time beginning a search of the top floor.

Truck personnel should not attempt to enter the building from the roof if the top floor is heavily charged with smoke. Instead, they should leave the roof and open the top-floor windows from aerial ladders or platforms. Ground ladders can be used to knock out the top-floor windows of shorter buildings. The top floor should be searched as soon as it is tenable.

The floor just above the fire floor must be thoroughly vented, for two reasons: first, that floor must be searched for victims; second, it must be checked for vertical fire extension, which must be controlled so the fire does not travel vertically to higher floors. The fire floor must be vented to allow the

Figure 5.5. Once on the roof, truck crews should use natural openings to begin venting, unless fire is directly under the roof. In that case, open over the fire first.

advancing of hoselines and a search for victims, as well as for fire fighter safety (Figure 5.6).

Ground-floor Stores

The stores on the ground floor of a combination occupancy are often structurally separated from the upper floors. That is, there may be no stairways or shafts extending up to the rest of the building from the stores. In spite

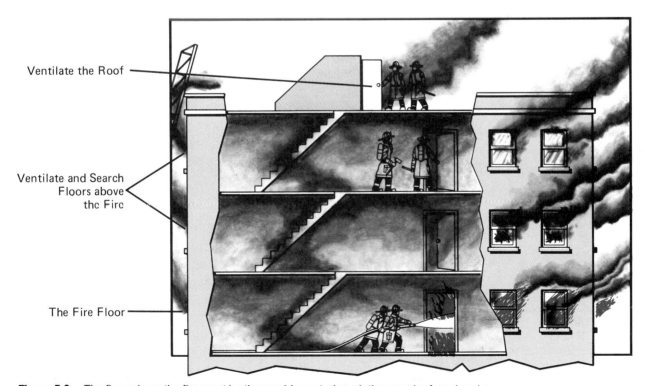

Figure 5.6. The floor above the fire must be thoroughly vented so victims can be found and checks made for fire extension. The same action should then be taken on the top floor.

of this separation, smoke and other combustion products will work their way up through the structure, especially if the store is heavily involved with fire.

It is necessary to ventilate the roof, the top floor and the floor above the fire, even when the fire is in a ground-floor store. The store itself should be thoroughly vented and the fire attacked quickly and aggressively. Utility shafts that serve the store and the upper floors, or other ground-floor stores, must be opened to deter fire spread.

Adjoining Buildings

When a multiple occupancy is well involved and is one of a row of similar structures, the adjoining buildings should be vented and their cocklofts checked for extending fire. This is especially important if the top floors of the fire building are involved or if there has been a backdraft or flashover at the top of the fire building.

An adjoining building should be vented through skylights, scuttles and penthouses to keep the damage to a minimum. However, if the cockloft is not vented when these features are opened, the boxed-in areas below them will have to be forced open. If the fire has extended into an adjoining building, this venting will release smoke and gases and slow their spread there.

The exposure fire should then be completely vented, with a roof opening at the hot spot if necessary. The fire must be reported to the officer in charge, who should order the operations necessary to contain the exposure fire.

SHOPPING CENTERS, ROW STORES, AND OTHER ONE-STORY BUILDINGS

Fires involving shopping centers, old row stores (sometimes called taxpayers), factories, warehouses and other large one-story structures sometimes cause excessive damage simply because they have not been ventilated properly. It is true that these buildings present only a limited problem in terms of vertical fire spread, but they are extremely vulnerable to horizontal fire spread.

Factories and warehouses usually contain large open areas through which fire can spread quickly. Shopping centers and row stores often have unbroken cocklofts and basements (possibly serving many stores), hanging ceilings, and unsealed openings in fire walls for utility pipes. All these features provide channels for horizontal fire spread from store to store.

Proper ventilation can stop or slow the spread of fire in one-story structures. Prefire inspection will help truck companies determine how the buildings in their territory should be vented.

Roof Operations

Truck companies arriving at a working fire in a large one-story structure should always assume that there are no fire walls between stores or building sections. This means that the roof should be opened for venting first. There is usually no problem in reaching the roof. A short ladder can be placed for access to the roof of an adjoining occupancy or to the roof of the involved building if it is wide enough. This keeps the ladders out of the way of engine companies advancing their attack lines.

A natural roof feature should be used for the first roof opening only if it is close enough to the hot spot; otherwise, the roof should be cut open at the hot spot or as close to it as possible. After one opening has been made over the

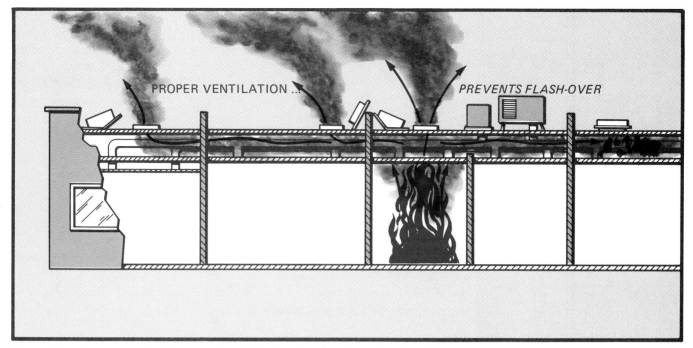

Figure 5.7. In row-type structures, horizontal fire travel will be stopped or slowed by proper venting.

main body of fire, natural openings can be used to complete the ventilation job (Figure 5.7).

Modern one-story warehouses are, in many cases, practically windowless. Roof venting operations must be effective to allow proper fire attack. These buildings often have several skylights, usually of the bubble type, installed across the roof. Those directly over the fire should be opened first, if possible (Figure 5.8). Then other skylights in the area should be opened. As in any other structure, if there are no natural openings above the fire, the roof should be opened at the hot spot.

Fire fighters must have a good understanding of roof construction and the effect of fire on the various roof types. The crews should be aware of any roofs in their area which might be particularly hazardous.

Figure 5.8. The roof should be opened as close over the fire as safety permits. If natural openings are located there, use them; if not, cut an opening. Natural openings should *not* be opened first if they are a good distance from the fire.

One word of caution: the steel roof girders and joists in many warehouses, garages and large one-story superstores are exposed from below, so there is no protective layer between the steel and the heat of the fire. Fire fighters reaching such a roof should check its condition carefully. If the roof feels spongy or is sagging, the steel could be warped and, therefore, weakened. Fire fighters should avoid such areas. Although they should open the roof as close to the hot spot as possible, they should not endanger themselves to do so.

Attached Occupancies

Once the roof of the fire building is properly opened, the roofs of attached buildings should be opened through natural roof features. This will allow truck crews to determine whether the fire has spread into the cocklofts of the attached occupancies (usually other stores). This also will allow accumulated gases to escape so they will not ignite when the ceilings of exposures are pulled down to check for fire extension. If fire is found in an attached exposure, its roof should be opened over the fire. Then its ceiling should be pulled down, and the fire attacked from below.

Ground-level Ventilation

When necessary, a store can be ventilated at ground level through its front display windows and through rear windows and doors. Some display windows are constructed with a large main window below, and window lights above. The window lights, being made of thinner glass than the main window, are easier to remove and cheaper to replace. They should be knocked out first. Since they are located near the ceiling (and near the greatest concentration of heat and smoke), they might provide sufficient venting. If not, then the main window should be removed.

BASEMENT FIRES IN LARGE STRUCTURES

A basement is the worst possible place for a working fire because it exposes the entire building and all its occupants. Since heat, smoke and gases will travel vertically throughout the building, spreading fire and overcoming occupants, it is imperative that the building be ventilated quickly and properly. The larger the involved area, the greater the ventilation effort required.

As a start, the basement, the first floor, and any shafts that lead up through the roof should be vented. The release of accumulated heat, smoke and gases from the basement and first floor will reduce the chance of vertical extension.

Basement Venting

Any opening into the basement, such as a door, window, chute cover or sidewalk door, can be used for venting. If possible, ventilation openings should be opposite those being used for fire attack. If the fire is being attacked from a rear basement entrance, then the basement should be vented from the front (and the sides if the building is detached). If the fire must be attacked from the front of the building, truck crews should ventilate at the rear. If a single available basement entrance must be used for fire attack, the area should be vented through available basement windows.

A basement with only one entrance (which is being used for fire attack) and no windows presents a serious venting problem. If the fire is extinguished quickly, venting the first floor may suffice to clear the basement. If not, the first floor should be cut through to open it to the basement just inside the first-floor windows. These windows should then be opened or knocked out to draw smoke and gas up from the basement (Figure 5.9).

Where basement windows do exist, they might be below grade and covered with steel grates. If the grate openings are wide enough to admit the end of a tool, the window glass can be knocked out through the grate. If not, the grate must be removed before the window is knocked out. For this, place a flathead axe (or the adz end of a halligan or kelly tool) between the grate and the concrete around it at the narrow end or at a corner. Drive in the tool and use it to pry up the grate.

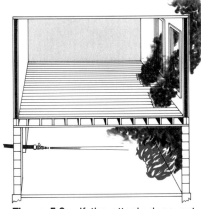

Figure 5.9. If the attack does not have immediate effect on the fire, the first floor should be opened just inside windows and the windows opened or removed. Fans can be used to blow the rising smoke out of the windows.

Storefront walls. When a basement fire is being attacked from the rear of the building or from an inside stairway, it is important that the front of the basement be ventilated. If there are no basement windows or doors at the front of the fire building, but the ground floor is occupied by a store, open the low wall below the display window to vent the basement. Often, the display-window wall opens directly to the basement, or to the basement ceiling, which can quickly be knocked down (Figure 5.10).

Before the wall is opened, the display window should be knocked out with a pike pole or short ladder. This helps make sure the display window does not break and shower glass on fire fighters while they are opening the wall. It also will help to vent the first floor.

Figure 5.10. Opening the wall under display windows can be an effective venting operation. First remove the glass, then the wall below the window. Check for a ceiling below that level.

With the display window removed, the wall can be opened. The procedure to be used depends on the materials involved. It might be necessary to cut through wood sheathing, pry steel panels off wood framing, or knock out brick veneer.

If the wall below the display window opens directly to the basement, smoke should pour out as soon as the wall is cut through. If not, there could be a basement ceiling between the basement and the wall. The ceiling can be pushed down with a pike pole or similar tool. The openings in the display-window wall and the basement ceiling should be made as large as possible for effective venting.

The display-window wall might be constructed of a material that is very difficult to remove, such as glass block. In such cases, the display-window floor can be opened to ventilate the basement. Again, the display window is first removed. Then the items in the window should be pushed back out of the way, and the display-window floor cut through near the front of the display area. Finally, the basement ceiling should be opened, if necessary. Smoke ejectors can be used to blow smoke out of the building as it issues from the hole. However, the ejectors should not be placed directly over the hole until the fire is knocked down (Figure 5.11).

When the fire is near the front of a long basement, and the store has two display windows, both display-window walls or floors should be opened. One opening can be used for fire attack with lines and the other for venting. It is usually necessary to vent the rear of the basement at the same time.

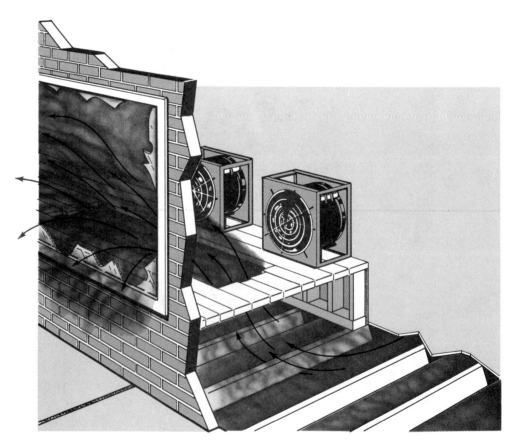

Figure 5.11. When ventilation is made through the display window floor, smoke ejectors can be used to blow the smoke out of the first floor.

Other openings. Any opening that will help ventilate the basement should be utilized to permit the advancement of attack lines and lessen the chance of vertical and lateral fire spread.

First-floor Venting

The first floor of the fire building should be vented through windows to remove combustion products that have seeped up from the basement. This will aid engine crews who might be advancing lines to the basement, through interior stairways, or to the first floor. The venting will also increase the effectiveness of search operations and reduce the chance of fire spread to the first floor.

Roof Venting

Shafts running up through the building must be opened at the roof to release combustion products convected up to that level. This will help vent the basement and will prevent smoke accumulation, ignition of the upper part of the building, and fire spread to the tops of attached buildings. If vertical shafts are not opened at a serious basement incident, fire fighters probably will be faced with a second fire at the top of the building.

FIRE RESISTANT STRUCTURES

Fire resistant construction is used in many structures classified as low-rise, medium-rise and high-rise. The main structural members of these buildings have special coatings to protect them from heat. Other construction features are designed to retain the heat of a fire in as small an area as possible. Doors to apartments and offices are made of steel, and most codes require that they be self-closing. Floors are concrete. Stairways and elevator shafts are enclosed. Utility shafts and heating and air conditioning ducts have high protective ratings. The interior stairways are constructed to serve as enclosed fire escapes.

Although these construction features restrict fire spread, they tend to result in extremely high temperatures in the immediate area of a fire and in very heavy accumulations of smoke and gases. A number of deaths and injuries can be traced to the spread of smoke and gases throughout these fire resistant buildings (by way of heating and cooling ducts), combined with lack of proper ventilation.

A fire resistant structure cannot be vented in the same way as a more standard structure — at least initially, when the building is occupied. For example, the enclosed stairways must not be opened to smoke-filled corridors while they are being used as escape routes. Therefore, they cannot be used for venting. To use an enclosed elevator shaft for venting, fire fighters would have to prop open elevator doors on the fire floor. This would be a definite hazard to fire fighters operating in a smokey corridor with minimum visibility. In addition, the fire might ignite grease and oil in the shaft and spread upward, or burning materials might fall down the shaft and ignite fires below.

Window Venting

The best way to ventilate an occupied fire resistant building is through its windows. To do so, truck crews must force entrance into the rooms, apartments, or offices on both sides of halls and corridors. The entry doors should be propped open and as many windows as possible opened in each unit. If the fire has control of a corridor, truck crews should advance along with the engine crews on the attack lines, force entry into a unit as they come to it, open the windows in the unit, and then rejoin the engine crews (Figure 5.12).

By opening units on both sides of the corridor, truck crews can create cross ventilation that will quickly begin to clear the area of combustion products. If necessary, smoke ejectors can be used to aid in the venting. As windows are opened, the fans should be placed to draw the smoke outside.

The fire floor should be vented first. As soon thereafter as possible (or at the same time, if personnel are available), the floor above the fire floor should be vented and searched for victims. Despite the fire resistant construction, smoke can seep up through the building. Engine companies should advance lines to the floor above the fire to keep it from extending vertically from window to window.

Stairway Venting

If the building is unoccupied or has been completely evacuated, doors leading from corridors to stairways can be opened. The stairway shafts will draw smoke from the corridors and provide vertical venting. In most cases, the penthouses above the stairway are fitted with ventilation openings, and truck crews need not climb to the roof to vent them. Such details should be determined through prefire inspection.

The larger the fire, the more pent-up heat and smoke will be encountered. Fire in a large open area in a fire resistant structure will require the use of heavy attack lines. Along with the smoke and heat, this will place great physical strain on engine company personnel, so truck crews must quickly remove as much heat and smoke as possible by venting through all available openings.

Other Truck Duties

It is extremely important in a fire resistant structure that the fire floor and the floor above it be vented, searched and checked for fire spread. In a building with a standpipe system, truck crews can use house lines to knock down extending fire until the arrival of engine company personnel. Truck crews must work closely with engine crews by ventilating to keep the hose lines moving.

In most fires in these structures, rescues will be made from the apartment, office or work area in which the fire originated. However, if the fire has spread into the corridors, or if smoke has begun to spread through some floors, frightened occupants — who would have done better to stay in their apartments or offices — might attempt to reach an exit but collapse en route. Therefore, while proceeding with ventilation assignments, truck crews must search every area through which they pass. This includes lobbies, elevator alcoves, corridors, the fire floor and the floor above it. Venting duties are secondary to the rescue of victims who have been overcome by smoke and gases.

USING ELEVATORS TO APPROACH THE FIRE FLOOR

If an elevator is available, it can be used to transport truck crews and their equipment to the vicinity of the fire floor in both fire resistant and standard

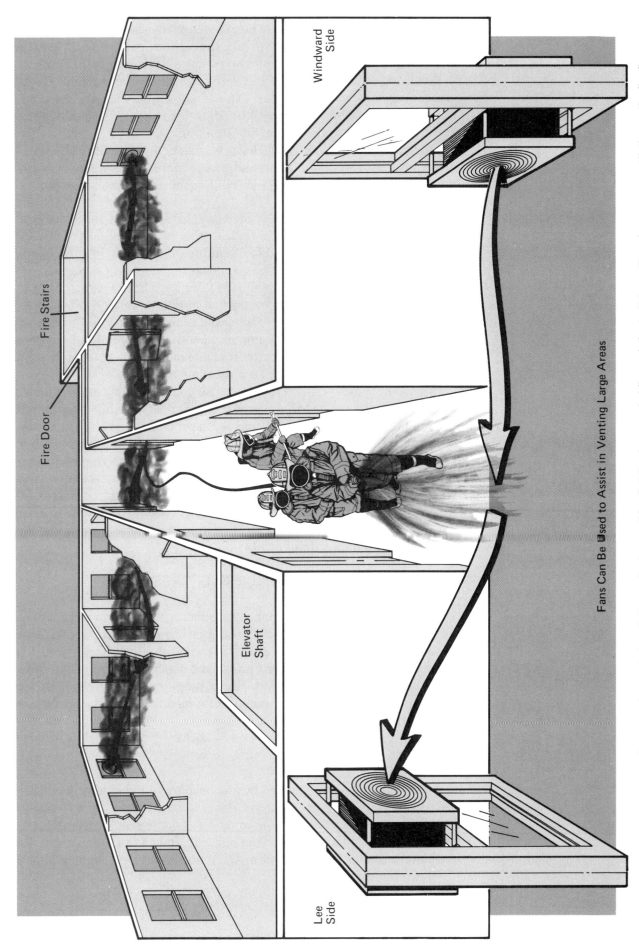

Figure 5.12. Vertical venting usually cannot be carried out in the early stages of fire fighting in occupied fire resistive buildings. Therefore, cross ventilation on the fire floor must be performed.

structures. However, the elevator should never be taken to the fire floor itself. If it is, those in the elevator might be exposed to flames or excessive heat when the door opens. It is much safer to take the elevator to the floor below the fire or, if the fire is intense, to get off two or more floors below the fire.

When the location of the fire is difficult to judge and might have been erroneously reported, fire fighters should get off the elevator three or four floors below the supposed fire floor, even if fire-sensing or smoke-sensing equipment is used to determine the fire floor.

Truck crews can also use the elevator as an equipment lift once personnel are in position. Tools, fans, lights and engine company equipment can be sent up from the ground floors, and an equipment pool established one or two floors below the fire.

Truck crews must become completely familiar with the elevators and elevator controls in buildings in their territory. Representatives of elevator companies should be consulted for accurate information on car-calling controls and their reaction to heat and moisture. In some cases, fire fighters intending to take the elevator to a lower floor have been taken directly to the fire floor by the action of heat on the call button there.

If the elevators have lobby controls, their locations should be known. Fire fighters should know how to obtain and keep control of elevators. Because elevators and controls vary greatly from one manufacturer to another, meetings with various manufacturers are recommended.

In an emergency, truck crews can use forcible-entry tools (which they should be carrying) to force elevator doors open or shut as required, to remove overhead escape panels, and — if necessary — to chop their way out of the elevator shaft.

In structures containing apartments or offices, it is often impossible for the fire department to obtain elevators immediately upon arrival since tenants will be using them to escape (despite warning notices to the contrary). In this case, if the building is not too tall, truck crews must use stairways to get up to the fire floor. In no case should self-contained breathing apparatus be left behind to lighten the equipment load.

SMOLDERING FIRES

On rare occasions, fire companies responding to an alarm will encounter a fire that is not, or does not appear to be, free burning. There will be plenty of smoke to indicate that there is a fire, but without visible flames. This condition exists because the fire is being deprived of sufficient oxygen to maintain open flames. In such a situation, fire fighters must assume that they have come upon a smoldering fire, which if not handled properly could become disastrous.

Indications

A smoldering fire is indicated by one or more of the following conditions:

- Much smoke is visible, but no open fire can be seen or heard.
- Smoke is rising rapidly from the building, indicating that it is hot. (Note, however, that humid weather or an atmospheric inversion may be holding down the smoke.)
- Smoke is leaving the building in puffs or at intervals.
- Some smoke is being drawn back into the building around windows, doors and eaves.

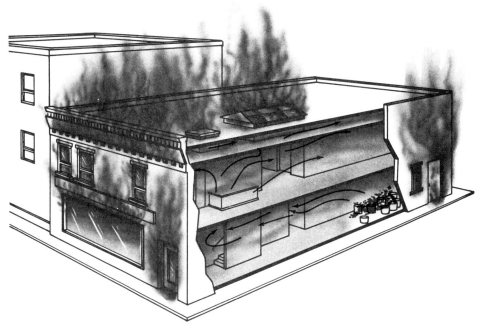

Figure 5.13. When indications of a smoldering fire are visible from the outside, the interior must be ventilated at the highest point possible before the building is entered.

- Although no flames are showing, windows are stained brown from the intense heat.

Occasionally, one or more window panes are broken by the heat inside the building. A small rim of fire can appear around the edges of the broken glass, where the oxygen content is high enough to ignite some of the gases. This indicates that a backdraft is imminent.

Whenever any of these conditions appears to be present, the fire must be handled as a smoldering fire for the safety of fire fighters and for proper fire fighting operations.

A smoldering fire has sufficient heat and fuel to become free burning. The heat comes from the fire, which was probably burning freely at one time. The fuel is mainly carbon monoxide gas from the original fire and from the smoldering fire; the contents of the building; and, perhaps, the building itself. The carbon monoxide has filled the building and surrounded the smoldering fire, thus cutting off its oxygen supply. Lack of oxygen keeps flames from developing, but the fire smolders and produces intense heat. A smoldering fire needs only oxygen to burst into flame, with no special circumstances involved. A fire can be smoldering in a building of any size or type or, in some cases, in only one area of a large structure (Figure 5.13).

Backdraft

A smoldering fire must be ventilated before it is attacked; that is, carbon monoxide must be removed from the building before air is allowed to enter. The addition of any oxygen to the heat and fuel will lead to immediate ignition.

This sudden ignition can take any of several forms. In one situation, the gases and heated combustibles might simply burst into flames, engulfing the building or a part of it in fire. In another, the force of the ignition might be

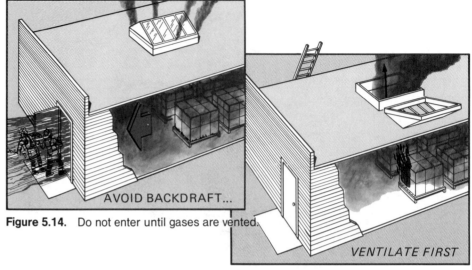

Figure 5.14. Do not enter until gases are vented.

Vent gases at highest point over the fire.

enough to blow windows, doors and fire fighters out of the building (Figure 5.14). There could also be an explosion strong enough to cause structural damage to the building and injury to fire fighters. Just what will happen cannot be determined beforehand, but it is certain that the addition of oxygen will cause some sort of ignition. That ignition is referred to as backdraft.

Venting

While the fire is smoldering, the carbon monoxide, other heated gases, hot air, and smoke will have been convected upward and will be collecting at the top of the structure. An opening must be made as high on the building as is safely possible to release these gases and allow them to move out of the structure (Figure 5.14). This is the same principle used to vent free-burning fires. The difference is that with smoldering fires the venting must be done *before* the building is entered, to relieve the explosive situation and reduce the chance of backdraft. If this sequence is not followed, air entering the building with fire fighters will cause immediate ignition.

It is important to ventilate fully and in the right places to ensure that the hot gases are dispersed. On one hand, ventilation should not be rushed or haphazard. On the other, it should be completed promptly. Heat, smoke and gases have been building up inside the structure, so it is possible for these combustion products themselves to break that window glass and allow air to enter before ventilation is completed. Again, the result would be sudden ignition; this time, however, fire fighters would be near the building.

Natural roof openings can be used to vent the building, and holes can be cut if necessary. Truck crews going to the roof will usually be able to determine which natural features to open first by noting the amount of smoke pushing out around them. If roof venting seems particularly dangerous, knock out the tops of the highest windows with an aerial ladder, ground ladders or solid streams. Once the top of the building is opened, the remainder can be vented as necessary.

Because a backdraft can take the form of a violent explosion, releasing a tremendous blast of fire and heat, fire fighters must avoid approaching the building directly. Their approach should be made either from an oblique angle or parallel to the building (Figure 5.15). This is especially important

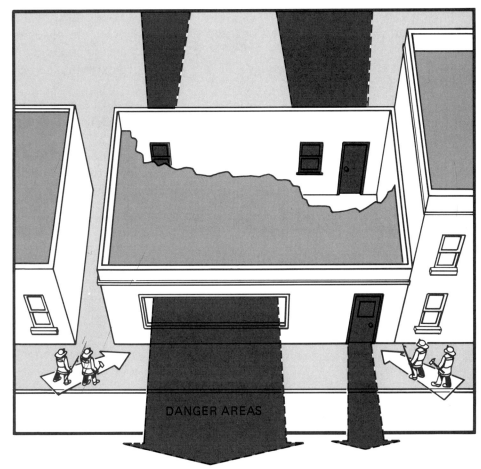

Figure 5.15. Approach to a building showing the warning signs of a backdraft should be from the side. Do not get directly in front of the building until venting has been accomplished.

when working at fires in stores or other structures with large glass areas, since these are the weakest points in the building. If a backdraft occurs, it will blast out through the glass. Many fire fighters caught in such a blast have suffered severe burns and critical injuries from flying glass and debris.

As a precaution, attack lines should be charged and ready for use during the ventilation of the building. Crews on the lines should be in safe positions, protected from flying glass, and ready to enter the building as soon as venting is completed. Likewise, if there is any possibility of a backdraft, apparatus should not be positioned in a direct line with the building, especially if there are large glass areas at the street level.

Once the fire is ventilated, it will burn freely. It can then be attacked in the same way as any other free-burning fire, although its size might not immediately be known.

SUMMARY

Every structural fire must be ventilated. A free-burning fire should be ventilated in coordination with other fire fighting operations. A smoldering fire must be ventilated as high as possible on the fire building before the building is entered.

In single-story and two-story dwellings, ventilation operations can consist solely of opening windows on the fire floor and on the floor above the fire, if there is one. In larger structures, truck crews might have to open the roof and the top floor, as well as the fire floor and the floor above it, in order to ventilate properly. In any building, the attic or cockloft must be checked for fire extension and vented if necessary. The tops of attached exposures should also be checked and vented.

Basement fires require quick and thorough venting of the basement, the ground floor, and any shafts that extend through the full height of the building.

Fire resistant buildings are vented mainly through windows on the fire floor and the floor above it. Because these buildings are designed to retain heat and smoke, truck crews must search for overcome victims as they perform their venting duties.

CHECKING FIRE EXTENSION 6

As a basic objective of fire fighting operations, exposure protection is second only to rescue. Exposures — structures or parts of structures not involved with a working fire but in danger of becoming involved — must be protected to minimize the danger to their occupants and to contain the fire. The major contribution of truck companies to exposure protection activities is to check carefully and thoroughly for fire spread.

Exposures are generally classified according to their location relative to the fire structure. Exterior, or outside, exposures are those that could become involved if the fire spread across some open area. Structures near, but detached from, the fire structure are exterior exposures; so are wings of the fire structure which are in danger of fire spread across an open courtyard or an air shaft.

Interior, or inside, exposures are those to which fire can spread from within the fire structure. They include floors in the fire building above the fire floor and structures attached to the fire building.

The danger to exterior exposures is usually obvious — flames leaping toward the structure, or burning embers being carried to it by convection currents and winds. Radiant heat itself cannot be seen, but it can be felt. Its effects on exposures are obvious; blistering paint, surface discoloration, cracking glass, visible vaporization (gas production), and ignition at various points are indications of the action of radiated heat.

Interior exposures are not so obvious. Truck crews must seek out hidden channels through which fire can spread. Although less spectacular than outside exposure fires, interior exposure fires can be much harder to extinguish.

As in every phase of fire fighting, area and building inspections and prefire planning are important parts of exposure protection. Inspections will help locate exposure hazards (conditions or situations in or around a building that could promote the spread of fire), as well as areas or neighborhoods in which the spread of fire is especially likely. Prefire planning should ensure that sufficient apparatus and personnel are dispatched on the first alarm to cover interior and exterior exposures.

When a fire obviously poses an extreme exposure hazard, first alarm truck company response should be increased over normal assignments. The additional fire fighters will be necessary so exposure protection operations can be carried out in the shortest time. In addition to their other fireground duties, truck crews will be required to locate secondary fires, to determine if fire is spreading through attached structures, to force entrance into exterior and interior exposures so that engine companies can attack spreading fire, and to open up interior channels through which the fire may be spreading. This last is perhaps the most important truck company duty in protecting exposures.

INTERIOR FIRE EXTENSION

Truck companies must check interior exposures to determine where the fire is located and to keep it from spreading to uninvolved areas. Fire can spread quickly within a building in almost any direction. Truck company personnel must move just as quickly to find spreading fire and, where necessary, to open up building features to check the extent of the fire and to provide access so engine crews can hit it with their streams.

The protection of interior exposures must be a coordinated effort between truck and engine companies. Truck crews check for spreading fire, and engine crews extinguish it. However, truck personnel need not wait for lines to be advanced before they begin to check fire extension. They may be able to get into position much more quickly than those on the lines; if so, they should begin checking immediately. In fact, engine crews on handlines might be assigned positions in the fire building as a result of the efforts of truck crews. If fire is found in a particular location, lines will be necessary there. If not, the lines can be better used elsewhere.

Fire fighters must check for fire extension up stairways and elevator shafts, through halls and corridors, and from room to room in an involved unit. If fire is found to be spreading in these places, there is a very good chance that it is also spreading vertically and horizontally through concealed spaces.

Fire in Concealed Spaces

There are signs to indicate that fire is spreading within a concealed space, but there are no signs to indicate that fire has *not* spread to a concealed space. If there is any possibility of fire in a horizontal or vertical space or shaft, it must be opened up and inspected visually. If necessary, streams must be directed into the shaft, and it must be ventilated.

Although this action will cause damage to the building, there is little choice in the matter. Either open up shafts, walls, partitions, ceilings, floors, or whatever, or let the fire destroy the building completely. While every effort should be made to minimize damage to the building and its contents, openings have to be large enough for inspection, hose manipulation, and ventilation. Openings must also be large enough to admit sufficient water to extinguish the fire.

The fact that fire will spread vertically and then horizontally until blocked, and that ventilation is important in controlling fire spread, was established in the last chapter. It follows that ventilation and the search for fire in concealed channels are companion operations. The opening of concealed spaces and ventilation of involved areas are truck work that must be done by truck company personnel in exposure protection operations.

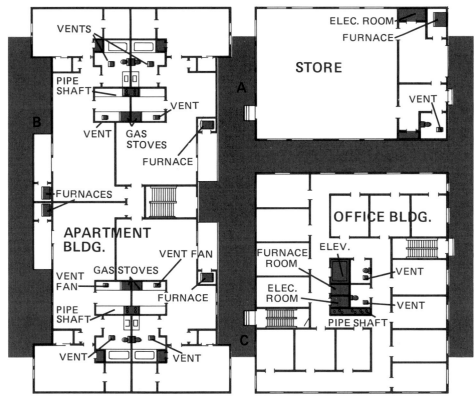

Figure 6.1. Locations of vertical channels vary with the occupancy type, size and age of the structure.

Vertical Fire Spread

Fire will travel vertically inside walls and partitions and through pipe shafts, dumbwaiters, air shafts, and similar pathways in a building. Many structures, including one- and two-family dwellings, contain concealed vertical shafts that carry water, gas and electric lines or sewer system vent pipes (Figure 6.1). Many of the newer single-family dwellings and air conditioned apartment houses have central-heating vents that extend up through the building to a chimney fixture on the roof.

These vertical channels are normally located toward the rear of commercial buildings, stores, and shopping centers (Figure 6.1A). In apartment buildings, they follow the pattern of the apartment layouts and are most often found near kitchens and bathrooms, each shaft usually located so it serves two, four or more apartment units (Figure 6.1B). In some modern structures, all the shafts are located in a single huge central core that runs from the basement up through the roof of the building (Figure 6.1C).

The great variety of designs for single-family dwellings means that vertical channels might be located almost anywhere in these homes. The locations of vent pipes and kitchen vents on the roof are good indicators of where these shafts will be found. Vertical concealed spaces are often created when the interior of a building is finished. Such spaces can be found under and around stairs, and within walls and partitions (Figure 6.2). These openings will rapidly spread fire up through the building.

Indications. If there is a working fire inside a building, fire fighters should assume that flames have entered concealed shafts until they determine otherwise. As they arrive on the fireground, truck company personnel should be looking for signs that fire has gotten into vertical channels within the build-

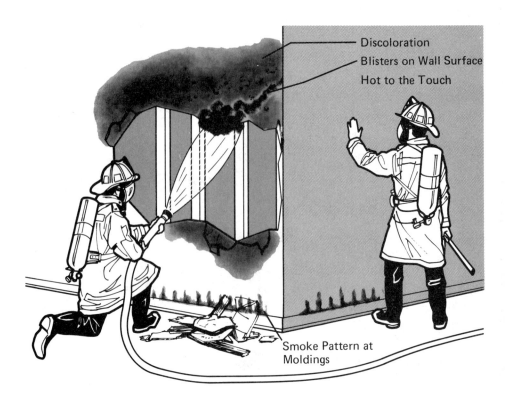

Figure 6.2. When extending fire is found in walls or partitions, lines should be stretched and the wall opened. The area must be thoroughly checked.

ing. External signs, mainly involving roof features and the upper parts of a building, were covered in the previous chapter.

Inside the building, truck crews should look for smoke and flames issuing from walls. Such signs as blistering, discoloration, or streaking of paint or other wall coverings indicate the presence of fire or heat within concealed shafts, walls or partitions (Figure 6.2). A wall that is hot to the touch is probably concealing fire. In the absence of other indications, fire might be detected by smell. Smoke within a concealed space gives off a strong odor that is easily detected. Fire fighters quickly learn to recognize the smell of burning electric wire coverings and wall insulating materials. At times, fire within a concealed space will actually crack, hiss and pop loudly enough to be detected by its sound.

Checking walls. The search for vertical fire spread should begin directly over the fire on the floor above the fire floor.

Baseboard areas should be felt for heat and examined for black streaks running up the walls. If either is found, the baseboard should be removed so the inside of the wall can be checked (Figure 6.3). If fire is extending upward within the wall, a line should be called for and the wall opened up to allow streams to hit the fire. If the fire is at the baseboard level only, or if there is heat but no fire, the wall need not be opened further. Streams should be directed at the opened baseboard area if fire is showing.

The walls themselves should be checked for the signs of fire discussed above. A wall that shows any of these signs must be opened to allow a stream to be directed onto the fire. The initial opening should be small and about waist high. If lines are available, the opening should be enlarged until a fire fighter on one knee can use the opening to direct a stream up into the wall. An

Figure 6.3. Check for vertical spread should begin directly over the fire. Baseboards can be removed to check extension through walls.

opening at waist level also allows the stream to be directed down into the wall with comparative ease.

A wall opening should not be enlarged unless there is a charged handline available with which to hit the fire. Once the line is in position, the opening should be further enlarged until the extent of the fire is determined and the fire is knocked down.

When fire is found to be extending up past the wall opening, the area above it must also be checked. This is true whether it means checking another story, an attic or a cockloft. If the fire has extended beyond that area, higher stories must be checked until the extent of the fire is found. This will require quick movement by truck and engine crews alike, but it is absolutely necessary. Streams must be directed onto any area that has been touched by the fire.

Checking vertical shafts. Truck crews on the floor above the fire must check all rooms that could contain utility shafts or pipes. These include kitchens, bathrooms, workshop areas, laboratories, janitors' closets and the like. Fire can spread vertically into these rooms through the shafts and other concealed spaces. In an apartment building, kitchens usually are located one above the other. Fire could enter the kitchen above the fire floor through a utility shaft and then spread up from floor to floor through that same shaft.

Built-in cabinets below the kitchen sink usually are constructed with a 3- to 5-inch enclosed space between the floor and the bottom shelf. Electric conduits, gas pipes, water pipes and drains often run through this enclosed space, so it is open to a wall or shaft. Fire in the space will travel to the wall and then up to higher stories if not found and extinguished quickly. Fire entering the space from below will travel horizontally through the space, as well as vertically (Figure 6.4).

Exhaust ducts in restaurants, hotels, hospitals and other buildings with large kitchens often develop a heavy internal coating of grease. If ignited, grease burns with a very hot flame that can heat the duct to the point where it, in turn, ignites combustible material placed against or even close to it. If the seams of the duct are loose, flames can push through them and ignite parts of the building. Thus, whenever a fire involves a grease duct or a room in which a grease duct is located, the entire length of the duct should be checked, all the way up to the roof.

Ductwork for forced-air heating systems and central air conditioning systems will, over the years, become matted with lint and dust. This material burns very fast once ignited, and the ducting system can quickly spread the fire throughout an entire building. Truck crews checking floors above the fire must check air intake and outlet registers for smoke, and the walls around

Figure 6.4. Built-in kitchen cabinets often conceal utility lines. When over a fire, these areas must be thoroughly checked.

ductwork for signs of heat. If either is found, a hole should be opened to determine the extent of the problem. If fire is found, a line should be brought into position, the wall opened, and the fire hit. Then, the rooms above must be checked in the same way.

In some air conditioning systems, the spaces between pairs of wall studs are used as return air ducts. This type of setup is most prevalent in single-family dwellings and garden apartments. The stud-space "duct" sometimes is lined with thin sheet metal, but often is not. In either case, since it provides a channel for the spread of fire up through the structure and into the walls themselves, stud-space ducts must be carefully checked for fire.

Pipe shafts might be completely concealed within walls, or located behind service doors or wall louvers on each floor of a building. The concealed shafts are obviously hard to find, as can be shafts behind doors or louvers which could be anywhere in the building. Unfortunately, these shafts are sometimes located in the individual units of hotel, motel or apartment buildings. In several notable cases, these shafts have contributed to loss of life when smoke, gases and fire were not released at the roof in time to keep them from pushing their way into the living units.

When such shafts are located by fire fighters, they should be checked for signs of fire travel. If any of these signs is found, a stream should be directed up the shaft through a forced opening, an open service door, or an opening from which the louvers have been removed (Figure 6.5).

When an intense fire is roaring up through a shaft, the floor and ceiling must be checked where they abut the shaft. If the floor is warm, or the ceiling shows any signs of fire, the area around the shaft must be opened and a stream directed into it. The quickest way to open the ceiling is to pull it down, starting with a small hole. If there is fire outside the shaft, the ceiling should be opened all around the shaft. As noted earlier, any shaft involved with fire should be opened at the roof.

Checking stairways. Fire can start in or find its way into storage rooms or framed-out (dead) spaces under stairways. The fire could then quickly spread to adjoining walls or shafts, or even into the ceiling, by traveling along the underside of the stairway (Figure 6.6).

Stairways and the spaces under them must be checked if they are near the fire, either on the fire floor or on the floor above the fire. Because the stairways will be used by escaping occupants as well as by advancing fire fighters, if at all possible the stairs must be kept intact and eliminated as a source of fire extension.

Checking doors and windows. Carpenters often leave a space between a door or window frame and the adjacent wall studs. Shims are placed in the space to level the frame and steady it, but the effect is the same as that of a vertical shaft. Fire will quickly extend up around the door or window.

The areas around doors and windows should be carefully checked if they have come in contact with fire or if fire has burned into doors and windows on the floor below.

Horizontal Fire Spread

Although the greatest tendency of fire is to travel vertically, it will also travel horizontally through any available paths. In the usual case, the fire starts in one area of a building and spreads up to the ceiling, where it is temporarily blocked from further vertical spread by the ceiling and floor

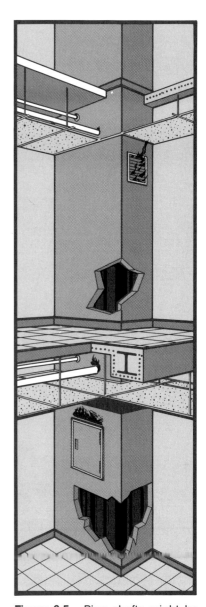

Figure 6.5. Pipe shafts might be completely concealed or located by noting the presence of service doors or louvers. In any case, they must be found and checked when in a fire area.

Figure 6.6. Dead spaces under stairs must be checked for spread of fire from below. If this is not done, fire will continue to spread vertically and possibly horizontally into ceiling space.

above. Eventually, it burns into the ceiling and walls, and then starts to travel through the building. If vertical and horizontal channels are available, the fire will spread through both, but it will travel faster through the vertical channels. If it is blocked from vertical spread, it will travel quickly through horizontal channels.

Fire can travel horizontally through the spaces between ceilings and floors, over false or hanging ceilings, through cocklofts, through and along ductwork and utility conduits, through conveyor tunnels in factories and warehouses, and through similar channels in other buildings. In addition, fire can spread through the concealed horizontal channels formed within walls, floors and ceilings by some types of construction (Figure 6.7).

Fire also can move horizontally between attached buildings or occupancies, through ducts, ceiling spaces and walls. An example would be the spread of fire from one adjoining apartment to another or through a row of stores.

All horizontal channels must be checked for signs of extending fire.

Indications. Truck crews arriving at a fire building will see few external signs of horizontal spread unless the fire has reached and involved the exterior walls. Inside the building, the signs of horizontal fire spread are the same as the signs of vertical spread. Truck crews must check floors and ceilings for smoke, fire, discoloration, hot spots, blistering paint, black heat streaks, and the sound and smell of fire.

Checking ceilings. When fire has control of an area, the ceilings of adjoining units — apartments, offices and stores in the fire building as well as in attached buildings — should be opened to determine if the fire is spreading. The fire is likely to spread horizontally from ceiling to ceiling.

Most ceilings are easy to open with a pike pole. First, a small slot should be opened along the common wall. If fire is found, a line should be called for. When it is in place, the ceiling should be opened up until the full extent of the fire is exposed and can be knocked down by the stream. If personnel numbers and fire conditions permit, salvage covers should be placed over furniture or stock in the area to keep water damage to a minimum (Figure 6.8). However, the most important objective is control of the fire.

Very high ceilings, such as those in old stores and in the lobbies of office buildings, are often difficult to reach and open with pike poles. Metal ceilings and those made of thick tongue-and-groove flooring material can take a long time to open. In such cases, it is probably best to open the flooring above the ceiling to check for horizontal fire spread.

On the other hand, ceilings consisting of square or rectangular tiles that lie on metal rails are very easy to open. The tiles are simply lifted off the rails with the tip of a pike pole. After the ceiling is checked and the fire extinguished, the tiles can be reset in the same manner. However, because the tiles are very light, they also can be lifted off the rails by the strong draft of a fire; and because the tiles are not sealed together, they allow fire and smoke to spread to adjoining units much more readily than do other types of ceilings. In many cases, the effects of fire on a tile ceiling have caused extensive smoke damage in units relatively removed from the fire itself. Where such tile ceilings are encountered, truck crews must be quick to check adjoining areas for fire and smoke.

When older buildings are remodeled, or when air conditioning or forced-air heating is added to a building, a new ceiling is often constructed below the original ceiling. Thus, there may be two or even three levels of ceilings in such buildings (Figure 6.9). After the lowest ceiling is opened, the others must also

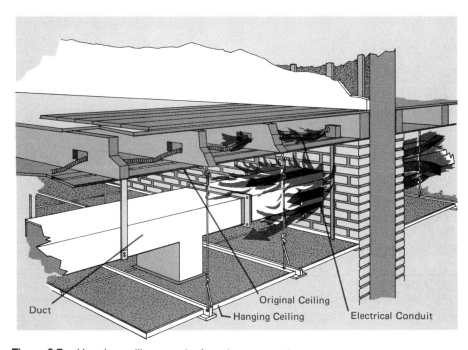

Figure 6.7. Hanging ceilings can be found over an entire store area or over certain counters that need special lighting.

Figure 6.8. A check for horizontal fire spread should be made on each side of the main body of fire.

be opened — whether or not fire is found above the lowest ceiling. If necessary, the topmost ceiling can be checked by opening the flooring or roof above it.

In some mercantile properties, mainly supermarkets, super drugstores, discount stores, and stores in older buildings, hanging ceilings cover all or part of the total area. Hanging ceilings might be located above the entire sales floor but not the stockrooms, or only above counters and display areas (in which case they usually were installed to lower the lighting fixtures). Hanging ceilings, sometimes used for storing light goods and empty cartons, contribute to the rapid horizontal spread of fire across the building and must be checked carefully.

Checking attached structures. The cocklofts or attics of structures attached to the fire structure must be checked for the lateral spread of fire. Truck crews must assume that there is nothing to stop the spread of fire through these spaces until they determine otherwise.

Proper ventilation will reduce lateral fire spread at the tops of exposed structures, as noted in Chapters 4 and 5, and venting crews on the roof might

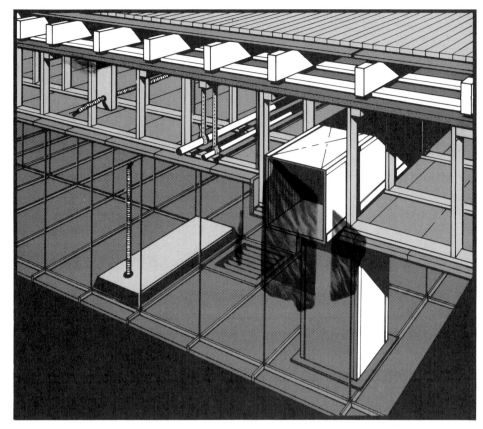

Figure 6.9. Alterations to the building might result in two or three hanging ceilings.

be able to detect fire spreading through attics. However, where there is any doubt, suspect areas should be checked through openings made in the ceiling below the attic or cockloft. In many fire situations, truck crews will be checking an exposure for interior fire spread before venting crews are sent to its roof. In these cases, the attic or cockloft must be carefully examined by the fastest means.

The basements of structures attached to the fire structure, especially in older buildings, also must be checked quickly. The party wall between two attached structures might support the floor joists of both buildings; there is often an opening in the wall where the joists overlap. Deterioration of the mortar over the years could have created a large opening through which fire can spread from one building to the other. In addition, large holes are often made in these party walls when plumbing or electric systems are modified or when new heating or air conditioning systems are installed. These holes are rarely closed up again, in spite of the fact that they will allow fire to spread into an uninvolved basement (Figure 6.10).

Open Interior Spread

A fire that has gained control of a large open area, such as a supermarket, warehouse or garage, can present engine companies with a serious problem. Truck companies can assist in deterring the lateral spread of fire by closing doors, windows, service openings and the like between the involved area and the remainder of the building. This can often be accomplished while truck crews are getting into position to check for fire spread, even before the attack lines have been advanced.

Fire doors are sometimes blocked open by occupants to ease the flow of

Figure 6.10. When a working basement fire is found in an attached structure, adjoining units must be checked for extension of fire.

foot traffic. "Automatic" fire doors, which are normally open but which close in the event of a fire, are sometimes kept from closing by stored materials. Truck personnel should make sure these doors are closed properly. (During prefire inspections, building occupants should be warned about the dangers of interfering with the operation of fire doors.) Generally, any opening between involved and uninvolved parts of the structure should be closed to slow the spread of fire.

EXTERIOR EXPOSURES

Truck crews can protect an exterior exposure from fire spread by entering the building and then closing windows and outside doors to keep sparks, embers and other burning material from entering. Fire fighters should remove curtains, drapes and shades from windows that face the fire to keep them from being ignited by radiated heat. They should check air shafts, narrow walkways or alleys, and open areas that are grown up with brush for fire spreading toward the exposure. They also should evacuate all occupants of the building.

If the exposed building has a wet standpipe system, truck crews should use the house line (the hose and nozzle attached to the outlet) to knock down fire that has entered the building. This is not recommended for engine companies, who should use their own lines, but truck crews would be foolish not to use any available means to protect the exposed building.

If the main fire is not controlled, or if exposure streams do not adequately protect the exposure, parts of the building could be ignited either by direct flame contact or by radiant heat. In this case, truck crews must search for victims if they have not had the chance to evacuate the building. They must also inform their officer or the officer in charge of the situation.

In general, truck companies operating in an exposed building should take whatever action is necessary to slow the spread of fire into or through the building. Once the exposure becomes involved with fire, it should be handled as a fire building rather than an exposure.

SUMMARY

Within a building, fire will travel through whatever paths are open to it — vertical or horizontal. The major vertical paths are walls and vertical shafts and ducts. Horizontal paths include the space between a ceiling and the floor above, as well as horizontal shafts and ducts. These are *hidden* paths, which must be opened by truck crews attempting to check for fire spread. When found, extending fire must be hit by engine crews using hoselines.

Truck crews can limit the spread of fire in large open areas and in exposed buildings by closing doors, windows and other openings between involved and uninvolved areas.

Some buildings, by virtue of their construction, are more prone than others to fire spread. Truck company personnel should be aware of these problem buildings and should be able to operate efficiently to check and control the spread of fire within them.

FORCIBLE ENTRY 7

To fight a building fire effectively, engine crews must advance their lines to the seat of the fire; truck crews must ventilate and check for fire extension, both in the fire building and in exposed buildings; and — most important — fire fighters must search for victims within these buildings. All these operations require that fire fighters be able to enter the fire building and exposed buildings.

One responsibility of truck companies is to provide access to locked-up structures, by force if necessary (Figure 7.1). Truck companies carry tools that can be used to cut, pry or force entry into a structure. They should also carry tools designed especially for forcible-entry work. Truck crews should know how and when to use these tools as well as which buildings in their territory will require forcible entry if they become involved with (or exposed to) fire.

This chapter deals with forcible-entry operations, from prefire inspection to fireground techniques, but it is not the purpose of this chapter to teach manipulative skills. For that, there are a number of commercially available skill manuals, visual aids, and training manuals to supplement company training sessions. Rather, this chapter deals with the efficient use of forcible-entry tools and skills in fire situations and with the effectiveness of forcible-entry operations.

PREFIRE INSPECTION

The greater the truck company's knowledge of its territory, the more efficiently it will operate on the fireground in forcing entry. As noted in previous chapters, prefire inspections improve the performance of all fireground duties. In preparation for forcible-entry operations, truck company personnel should determine through prefire inspections:

- Which buildings are locked up during part of the day, and the times at which they are locked (and so would need to be forced open).

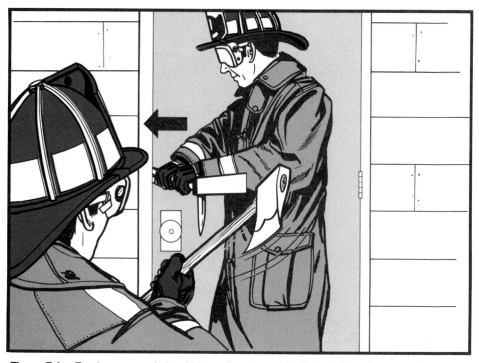

Figure 7.1. Truck personnel should know how to use their tools in order to perform quick forcible-entry work.

- Which buildings are always open at the street entrance, but could require forced entry into individual units, e.g., most apartment buildings, hotels and motels, and some office buildings.
- Which buildings are locked at a street entrance and at an inner lobby door, both of which might have to be forced.
- Which buildings have doors that, when locked, can be easily forced open, and which are difficult to force (for the latter, truck companies might be able to obtain keys to be carried on the apparatus).
- Which buildings can be entered from the rear and the sides, as well as from the front. Normally, front entrances are the easiest to force, but the locations and construction of windows and doors at the sides and rear might allow them to be used most effectively for forced entry — especially if the front door is difficult to force.
- Which buildings have private security forces that will respond to an alarm with keys, thereby eliminating the need for forcible entry.
- Which buildings might present forcible-entry problems as exposures if a nearby or attached structure becomes involved with fire.
- Which is the best way to enter problem buildings by force if that should become necessary.

The results of such inspections can often be used in planning truck company operations. One example would be a case in which keys for the fire building are carried on the truck; then it would probably not be necessary to assign fire fighters to forcible-entry duty on arrival.

Results of prefire inspections might also be of use in pointing up the need for special forcible-entry tools in a particular territory, in positioning truck tools on the apparatus (so that the most-used tools will be easiest to reach), and in assigning front and rear coverage and exposure coverage to first-alarm truck companies.

Such prefire inspection should be a continuing effort, since most building owners are constantly seeking to improve the security of their structures. As much as possible, truck companies should be aware of changes in the way buildings within their territory are locked. Changes made to increase the difficulty of unauthorized entry also increase the difficulty of entering a fire building. Although it probably is impossible to know everything about a territory, truck crews should certainly be aware of how best to enter those buildings with unusual or extremely difficult entry problems.

SIZE-UP

Forcible entry is another truck company operation that appears to add damage to the fire building. However, the small amount of damage done through forcible entry allows fire fighters to get into position quickly, can result in the saving of lives, and greatly reduces overall damage. The need for forcible entry will be indicated by prefire inspections, along with initial size-up of the situation (including the type of occupancy, the rescue problem, and the location and extent of the fire).

The Fire Building

First-arriving truck companies might find little or no sign of fire, a working fire, or a smoldering fire in a building that must be entered forcibly. If there are no signs of fire, the building can be checked quickly to determine the easiest way to force entry. In this situation, truck crews have the time to force entry carefully, so they do the minimum amount of damage.

When a working fire has gained headway — especially when it threatens to cut off escape paths or has trapped occupants — arriving truck companies must work quickly and decisively. They must force entry immediately without stopping to consider the damage they might do. The faster the building is opened, the sooner the building can be searched, the fire attacked, and the combustion products vented. A working fire justifies quick entry by the most expeditious means.

When arriving truck companies find (or suspect) a smoldering fire, they must not enter the building until it has been properly ventilated.

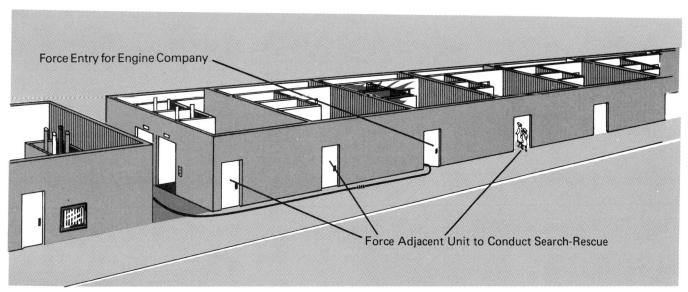

Figure 7.2. In apartments or office buildings, units on both sides of fire must be opened for search and advancement of hoselines.

Once inside the fire building, truck personnel might have to force entry to individual units in order to perform other truck duties, especially to search for victims (Figure 7.2). They also might have to force doors within the fire building to ensure that they will have access to units or parts of the building for later fire fighting operations as needed. This will depend on the size and location of the fire, the type of occupancy, and the locations of locked doors relative to the fire.

Exposed Buildings

When the fire building is one of a row of attached structures, truck crews should force entry into the building on each side of the fire building. They also might have to force entry into exposed structures that are taller than the fire building — either detached but nearby, or attached to the fire building (Figure 7.3). Whether or not truck company personnel have to enter these buildings at the time they are opened, they should be opened so that fire fighters can quickly enter them if necessary.

The object of forcing entry into exposed buildings (and exposed parts of the fire building) is to provide access ahead of time — that is, to stay ahead of the fire — so that no time will be wasted in entering the exposure if and when this action becomes necessary. Truck company members must check to be sure they have provided access to all parts of the building. They might have to force inside corridor doors as well as the front (street) door and, perhaps, a lobby door.

Figure 7.3. When detached buildings are exposed to the fire, their entrances should be forced so their interiors can be reached quickly if necessary.

Truck companies should not force entry into exposures when force is not required. For example, if the exposure is a residential occupancy, the occupant or manager might be on hand with a key. A quick check among bystanders for such a person might eliminate the need for forcible entry. In some cases, a cleaning crew working in an exposed office building can be quickly summoned to open the building. Always check the entry door before it is forced, since it may not have been locked.

TOOLS

"Forcible entry" implies speed; a building is forced open so fire fighters can enter it quickly. The forcible-entry operation itself should be carried out

Figure 7.4. Eye and hand protection is a "must" when using any tools, especially power tools. Crews not using tools should stand clear of the work area.

as quickly as possible and should create as little damage as possible. Both speed and minimal damage are achieved through proficiency with forcible-entry tools — a proficiency that comes only as a result of proper and continual training.

A truck crew that is well trained in the use of forcible-entry tools will work quickly and efficiently with minimal damage. A poorly trained crew will work slowly and will not use the tools correctly. Incorrect use of the tools will result in excessive damage, even if the crew works at a snail's pace.

Personal safety must be emphasized — especially with regard to power tools — in training sessions as well as in actual operations. Eye and hand protection are of the utmost importance. A truck company member who neglects this protection to save time will defeat the purpose if an injury results. Truck personnel must be extremely careful when working with hand tools near glass, when using power saws and explosive tools, and when using air- or hydraulic-powered forcing tools. Anyone not actually involved in the operation should stand clear of the area (Figure 7.4).

Cutting Tools

The cutting tools most often used for forcible entry are pickhead and flathead axes, bolt cutters, power saws, and air-operated and hydraulic cutters. In addition, the adz (chisel) end of a halligan or kelly tool can be used to cut and can be driven by a flathead axe or maul. Of the two axes, the pickhead is usually kept sharper for cutting. The flathead axe, not sharpened to as fine an edge, is useful for forcing and prying.

Some departments use cutting torches for special entry problems. Torches are particularly effective in cutting bars away from windows and doors and in cutting through roll-up metal doors, but care must be exercised to make sure the torch does not start another fire.

Prying and Forcing Tools

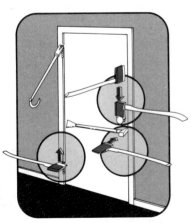

Figure 7.5. Prying and forcing tools can be used singly or in combination.

The tools available for prying and forcing work are the halligan tool, claw tool, kelly tool, pry-axe, hux bar, and similar devices. The flathead axe is often used for such work, either alone or driven by another flathead axe, or to drive some other tool (Figure 7.5). For heavy work, a maul or hammerhead pick can be used to drive another tool, or the maul can be used to drive the pick.

Hydraulic- and air-powered forcible-entry tools are available in various sizes and types. Often such tools are purchased for nonfire rescue work and then overlooked on the fireground. Depending on their design and capabilities, these power tools are used to force doors open, to raise roll-up doors, to remove or spread bars over windows and doors, and for similar prying and forcing applications. In an area where heavy or barred doors are common, power tools should be standard truck company equipment, and truck company personnel should be well trained in their use.

The battering ram has been shown to be effective for breaking through heavy doors and forcing openings through walls. Unfortunately, this device is no longer a required item of truck company equipment, even though increased building security measures make it more useful than ever. Fire departments should order a battering ram along with new apparatus or should bring present equipment up to date.

Lock Pullers

Lock pullers, such as the K-tool, are designed to remove cylinder locks. When operated properly, they do this quickly and are especially useful in opening steel doors equipped with cylinder locks. A truck company with many stores, dwellings and apartment buildings in its territory should be equipped with at least one lock puller. In use, one part of the lock puller is driven onto the cylinder lock and then pried off with a halligan or similar tool. The lock is pulled out with the tool, then another part of the tool is used to release the latch.

Explosive Tools

Explosive tools for forcible-entry work are sold under various brand names. Essentially, they are shaped charges which, when triggered, exert all their force in one direction. They can be used to blow a hole in a door, blow out its lock, or otherwise aid truck crews in getting doors open. They are very effective against steel doors and other hard-to-open entryways. Explosive tools are also available in sizes that will quickly open a hole in a roof for ventilation. They must, however, be used in strict accordance with manufacturers' safety and operational directions, since they can be dangerous when used improperly.

FORCIBLE ENTRY THROUGH WINDOWS

As noted earlier, the way in which truck company members force entry to a building depends in part on the fire situation. If a working fire has control of a large part of the building or has trapped some occupants, entry must be forced the quickest way. This is usually through a ground-level door or window, which might have to be cut, pried or forced open without regard for damage.

When no fire is encountered by an arriving truck company, the situation is not severe. Truck personnel can make a quick examination of the building to look for the best place to force entry. In this situation, the best place is often a window above the ground floor.

Thus, at its simplest, forcible entry may require only that one fire fighter climb a ladder, open an unlocked second-floor window, and then open the street door from inside. By contrast, the most serious forcible-entry problem might require cutting through a steel door or ramming through a wall. Forcible entry through windows is discussed in the remainder of this section, and through doors and walls in the next section.

It is important to note that the way in which the forcible-entry operation is implemented also depends on the occupancy, especially the number of people who could be in the building. This, in turn, usually depends on the time of day and the day of the week. The type of occupancy, its construction, and its population at the time of the alarm can reduce the options of truck crews in forcing entrance to the building. Such information should be known before the alarm, through prefire inspection.

During a working fire, fire fighters usually gain access to a building through a ground-level door or window. However, street-level windows in some buildings are barred, shuttered or otherwise fortified against burglary and vandalism. This is especially true of rear windows in commercial buildings. When the fire situation requires that truck crews force entry through such a window, they should use one above the ground level. The window is reached by a fire escape or ladder.

If a window at or above the ground level is unlocked, there is no entry problem. If the window is locked, it must be forced open.

Double-hung Windows

The window that allows the simplest and quickest access to a building is the large double-hung window. This type can be forced without much strain by prying up the bottom section at the center of the window. If the top section is made of small panes, the pane nearest the lock can be removed and the window unlocked (Figure 7.6).

If a double-hung window that must be used for entry cannot be forced quickly, the window should be completely knocked out. If it is at ground level, this can be done with an axe or another appropriate tool. Above ground level, the situation might not be discovered until a truck company member reaches the window by ladder. Then, the window should be removed with a tool. If it is safe to enter, the fire fighter should remove all splinters of glass and frame before moving through the window.

When a ladder is placed to a window for entry work, it should be positioned upwind if the wind is a factor. Any smoke, gases or fire coming out of the window will be carried away from the fire fighters without endangering their lives or impairing their vision.

When time and/or the fire situation do not permit these actions, the window should be quickly knocked out with the top of a ladder. Then the ladder should be placed as noted above. The first fire fighter up should use a tool to remove any remaining shards and splinters before entering.

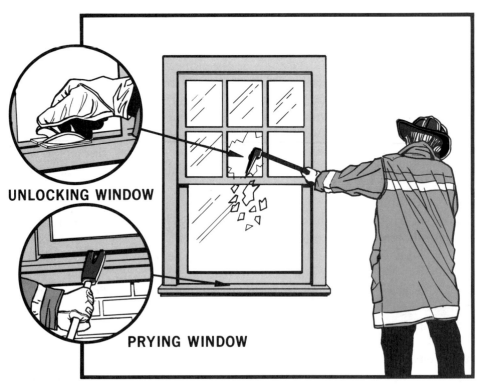

Figure 7.6. Double-hung windows can be opened by removing glass and unlocking, or by forcing.

In high crime and vandalism areas, the glass panes of lower floor windows are sometimes replaced with unbreakable plastic panes (such as Lexan®). Although this type of material can be cut with a power saw, the quickest means of entry (with other than steel window frames) is to knock out the entire window frame. In some cases, knock-out panels will have been installed in the window; these can be removed by striking a corner of each panel with the pick of an axe.

Casement Windows

A casement window also will allow easy access, provided that the opening is large enough to admit fire fighters. A casement window is hinged vertically, with the moving part of the window attached to a crank. The cranking assembly is usually light, but the window lock, usually located at the middle or bottom of the window, might be quite strong.

The best way to open a large casement window is to break out a pane of glass, reach in and unlock the window, and then force it open with a pry tool (Figure 7.7). The cranking assembly should give way without much resistance. If the heat is not intense, there might be time to remove a second pane, reach in, and crank the window open.

Many casement windows are too narrow to allow entry. However, such narrow windows are often located at the sides of a large glass picture window, which can be knocked out if the casement sections are troublesome.

Other Windows

The design of some windows prevents their use for quick access. They may have very heavy metal frames, wire within the glass, horizontally hinged sections that swing out when the window is opened, center swing-out sections surrounded by stationary panes, and so on. Some windows may simply be too small to allow entry. Truck companies should be aware of the locations of such windows in their territory (again, through prefire inspection) and should

Figure 7.7. When a casement window is large enough to admit fire fighters, it should be opened by knocking out a pane, releasing the lock, and forcing it open.

not expect to use them for forced entry. Alternative entryways should be determined in such cases.

Large windows, such as picture windows, of the double-pane type are expensive to replace. If such windows have been damaged by heat or smoke, there should be no hesitation in removing them; but rather than forcing an undamaged double-pane window, some other quick and safe entry should be sought.

Storm windows or screens must often be removed before built-in windows can be opened. In many cases, hooks or holding clips can be undone to allow these units to be removed quickly with little or no damage. However, if this cannot be accomplished quickly at a working fire, the storm windows or screens should be knocked out.

FORCIBLE ENTRY THROUGH DOORWAYS

When the fire situation demands that fire fighters quickly enter a locked-up building, truck crews will probably be assigned to force entry through a doorway. Although there are a number of different types of doors, each type is more or less associated with a particular occupancy. For example, heavy steel doors can be found at the rear of a commercial structure, but rarely on dwellings. Knowing the area or the particular building to which they are responding, truck companies should therefore have some idea of the type of entry problem they might encounter. They should, however, be prepared to force entry into every type of occupancy in their territory.

Commercial Occupancies: Front

It is almost always easier to force entry through the front door than through the rear door of a store or other business establishment. In an older building, the front door might be constructed entirely of wood, or of a wood frame surrounding ordinary plate glass. In more modern structures, the front door is often made of tempered (and virtually unbreakable) glass, with or without a frame, or of heavy plate glass in a strong frame. One, two or more such doors can be set into the doorway. Rear doors are usually made of steel or reinforced with steel.

The front doors and display windows of many business occupancies are further protected by metal shutters, accordion-type barred gratings, or similar devices installed to prevent vandalism and burglary. However, these devices also hinder fire fighting operations. Such a device must be forced open before the door or display window can be opened. Most of these protective devices are secured with a padlock that can be forced open quickly with a halligan or claw tool. Others, locked into a device set in concrete, take longer to force. Whichever way they are locked, the devices must be opened so fire fighters can get to the doorway.

Tempered-glass doors. For all practical purposes, tempered glass cannot be broken, and it would be a waste of time to try. Instead, a tempered-glass door should be attacked at its lock, or some other means of entry should be found.

The locks on tempered-glass doors are usually cylinder locks, located at the middle or at the bottom of the door. Double tempered-glass doors are usually locked at the middle. In either case, a lock puller such as a K-tool can be used to remove the lock. When no lock puller is available, the adz end of a halligan or kelly tool can be used. If the lock is at the middle of the door, the tool is driven in between the door and frame (or between the two doors of a double door) until the lock is forced open. In an alternative procedure, the tool is first driven into the space above the lock and then driven down to destroy the locking pins. If the lock is at the bottom of the door, the claw end of the tool is driven in under the door until the lock is raised out of its keeper. This method will work with any type of lock located at the bottom of a door. Also, hydraulic tools such as a small porta-power or rabbit tool can be used to force apart double doors or to raise doors locked at the bottom.

In many front entrances, tempered-glass doors are set between stationary plate-glass panels. The panels are usually as high as the door and wide enough to admit fire fighters. The quickest way to force entry through such a doorway is to break out the plate glass, leaving the tempered-glass doors in place (Figure 7.8).

Many stores, especially larger ones, have one or more display windows near the tempered-glass entrance doors. Again, the doors should be left in place and the building entered by knocking out the display windows (as discussed in Chapter 3). This is especially effective where the display windows open directly to the sales floor, so they can be used for advancing lines, venting and other fire fighting operations.

If a tempered plate-glass door must be broken, it is best to do so by striking a lower corner of the door with the pick end of an axe. When the door breaks, it will shatter into hundreds of pieces — not sharp, but more like globules. However, the fire fighter attempting to force the door should get as much of his back toward the door as possible when using the axe. Other fire fighters should stand back away from the door and axe.

Heavy plate-glass doors. Doors with heavy plate glass set in heavy frames should be treated in the same way as tempered-glass doors. Although it

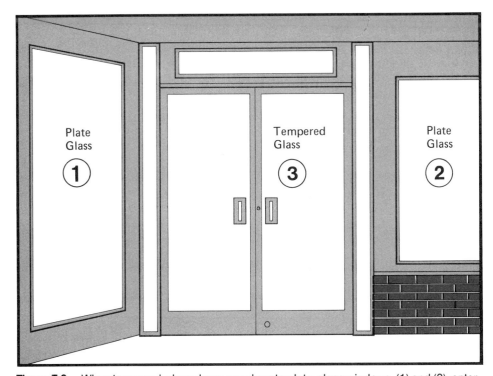

Figure 7.8. When tempered-glass doors are close to plate-glass windows, (1) and (2), enter through these windows. If entry must be made through door, (3), force or pull the lock.

Figure 7.9. Where there are older type doors with plate-glass panes, it is generally quickest to remove the glass and open the door from the inside.

might be possible to remove the glass quickly, there is usually a bar across the center or lower center of the door. The bar would have to be torn off the door before fire fighters could easily get through the opening. In addition, the glass could shatter and fly at fire fighters if the door is pushed too far out of shape while being forced. It is much better to remove or force the lock or to enter through a nearby plate-glass window.

Wooden doors. Wooden doors might or might not have cylinder locks, but they usually have bolts that engage keepers at the top or bottom of the doorway, or at both places. Double wooden doors can be bolted to each other (Figure 7.9), so pulling or forcing the lock does not guarantee entry. However, these doors usually have center panes (or panels) that can be quickly broken out. If the opening is sufficient for entry and exit, the door can be left in place. Otherwise, one fire fighter can reach through the opening and unlock or unbolt the door from the inside.

Commercial Occupancies: Rear

Commercial occupancies usually can (and should) be entered at the front. Rear entry is very often complicated by steel doors, barred doors and windows, and roll-up doors. There are seldom any display windows at the rear. However, where a rear door is constructed like a front door, it should be forced in the same way.

Steel doors. Before an attempt is made to force a steel door, it should be checked for an exposed lock or exposed hinges. If the lock can be seen, drive in a pry tool between the door and frame, and force the door open. If the hinges are exposed, pull the hinge pins and open the door from the hinge side (Figure 7.10), or drive a tool between each hinge and the door facing to force the hinges loose.

Doors with neither a lock nor hinges exposed cannot be forced with standard tools. Such doors might be secured with a steel bar placed across the door and door frame and held in place by heavy steel hangers. They might be

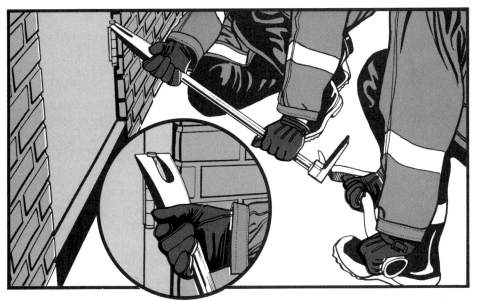

Figure 7.10. When hinges are exposed on steel doors, the quickest entry is usually to remove the pins and force the door from the hinge side.

locked with a fox lock — a device with from two to eight bars that hold the door closed from the inside. In a fox lock, the bars are attached to a rotating plate on the door. The plate is rotated one way to move the bars into keepers in the door frame, and the opposite way to withdraw the bars and allow the door to be opened.

A steel door that cannot be forced can be cut open (or its lock cut out) with a power saw or a cutting torch. A heavy steel door can be opened with a battering ram. The ramming will force the locks from their keepers, bend the door enough to pull the locks from their keepers, or tear the keepers out of the wall or door frame.

A steel door equipped with a fox lock is practically impossible to force. The operation is so time-consuming and requires so much effort that by the time the door is finally opened it is usually no longer effective for fire fighting operations. If possible, an alternative entry should be found. If a steel door with a fox lock must be forced, an explosive shaped-charge type tool should be used.

Breaching walls. It is sometimes quicker to open a wall than to force a steel door, especially if the wall is made of concrete block or cinder block. Mauls, battering rams and hammerhead picks can be used to make openings in most walls, including brick; block walls can be cut with power saws. An aggressive attack often will open a wall in a short time.

If possible, the wall should be opened near the doorway. The door and door frame will help support the opened wall, and the opening will lead to a corridor or other open area. At first, the opening should be made only large enough to permit streams to be directed inside to knock down fire and cool the interior. Then the hole can be enlarged enough to allow access for fire fighters. At this point, truck crews must be sure that blocks or bricks over the opening are firmly in place. If possible, the enlarged opening should have a keystone shape.

Roll-up doors. Doors that open upward might be locked in any one of several ways. Some, usually wooden, are locked with a modified fox lock. The lock and a handle are located at the center of the door, and the bars engage keepers at each side. These can usually be opened quickly by knocking out a door panel near the lock and reaching in to rotate the lock handle.

A wooden roll-up door might be further secured with pins that extend from the sides of the door into the track on which the door rides. The door should be pried up from the bottom to bend the pins out of place. In some cases, a ring on the door is padlocked to a ring set into the floor. This arrangement can be opened by forcing the claw end of a halligan or claw tool under the door against the rings and driving the tool with a flathead axe or sledge (Figure 7.11).

If a wooden roll-up door is difficult to force open, it can be cut with a power saw or axes. When cutting or prying these doors, truck crews must be careful of glass panes set in the doors. If the glass breaks while a door is being forced, shards will fly in all directions.

Metal roll-up doors do not usually have built-in locks. They can be padlocked to the floor or locked into their rails with pins. Manually operated doors are often locked through the chain used to raise and lower them. A motorized door is rigidly connected to its operating mechanism.

The first step in forcing a metal roll-up door is to pry it up as much as possible at both sides. If the door is found to be locked to the floor with rings and a padlock, the locking assembly should be forced as described above. If no locking rings are found, the door might be locked with pins or through its chain. Continued prying should bend the locking pins out of the rails or separate the chain. A hydraulic spreader can be of help here.

If only one side of the door can be raised, it might be possible to provide adequate access by wedging up that side. In some cases, one fire fighter might be able to get inside, under the door, and then release the latches to get the door opened. If nothing seems to work and the door must be opened, a hole should be cut in the door with a power saw, torch, or explosive shaped-charge.

Light doors. In many older buildings, rear doors are made of wood or light metal, reinforced with bars or fitted with several locks. This is often the case in older row structures with small shops on the ground floor and apartments or offices above.

The main (built-in) lock of such a door should be forced first. If it can be sprung, the additional bolts and locks can usually be forced with hand tools and brute strength. The various pry tools and the flathead axe will be most effective; the axe can be used to drive other tools or to pry directly on the door.

If the door has a glass pane without bars, it is best to remove the glass and attempt to open the locks from the inside — provided the location of the fire does not prevent such action. As always, fire fighters must use caution when attempting to force a door that contains glass.

Dwellings and Apartments

Locked residential structures are, in general, more easily entered than commercial structures. Front and rear doors are usually of the same type and of light construction. They often have one or more glass panes that can be knocked out, allowing truck crews to open the locks from inside.

The hand tools carried on the apparatus are more than adequate to quickly force entrance into a locked one- or two-family dwelling.

Truck personnel usually need not force entry into a multiple-unit residence

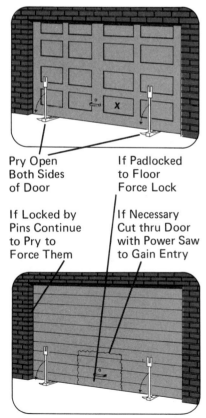

Figure 7.11. Forcing of roll-up doors depends on their construction and locking features.

since the street doors at the front are often unlocked. The rear doors might be locked during the night. There may be a lobby door, secured by an electric lock that can be tripped from each apartment. In older buildings, these doors are light — usually of wood and glass — and can easily be forced. Newer apartment buildings might have tempered-glass lobby doors, with plate glass at each side. These doors can usually be forced at the electric lock, or the plate glass can be knocked out.

In some modern high-rise and garden apartment buildings, the lobby doors and the doors to interior stairways are usually unlocked. However, if they are found to be locked, they can be forced quickly by methods described earlier in this section.

Apartment doors. Once inside an apartment building, truck crews might have to open individual apartment doors for search and rescue, venting, and to check for fire spread. In older buildings, apartment doors are usually made of wood; cylinder locks might have been added, and the original door locks retained. Cylinder locks can be pulled with a lock puller, and the doors forced with any prying tool. The frames of these doors are usually strong enough to support a pry tool, allowing doors to be forced fairly easily.

In more modern apartment buildings, apartment doors are made of steel or of wood covered with steel. Most often, they are secured with cylinder locks and possibly one or more bolt-type locks. Here again, a lock puller should be used to begin the forced entry; or a handheld pry tool should be placed just above or just below the lock, driven in with a flathead axe, and then grasped at the outer end and pulled away from the door. Use of the porta-power or rabbit tool (Fig. 2.1) is also effective for quick opening of these doors.

In some cases the hydraulic type smoke ejector hanger can be used to force a door. Place it across the door just above the lock and extend it until it allows the door tab to clear the striker plate.

If the door frame is constructed of light metal, it might not support the pry tool. In such a case, the tool should be driven in further than normal, and the door struck with the back of a flathead axe as the end of the tool is pulled away from the door. This will usually spring the door loose. The pressure exerted in forcibly opening an apartment door will usually tear loose or break any chain lock.

Floor locks (police locks). In high-crime areas, truck companies might encounter the floor lock, or brace (Figure 7.12). This device consists of a heavy bar fastened to the floor inside the apartment door and to a plate on the door. The bar can be pivoted away from the door when the tenant opens the door, and it slides into a plate mounted on the door for locking.

A floor lock is very difficult to force. A wooden door can sometimes be pulled away from the device at the hinge side, or it might be possible to cut the door open with axes to allow someone to reach in and release the floor lock. However, in some of these devices, the bar is locked to the plate in the door, and the door must literally be destroyed before the apartment can be entered.

The combination of a floor lock and a steel or steel-covered wooden door presents an even greater problem. It is extremely difficult to break through the door or to open it at the hinge side. To add to the problem, a police lock is usually used in combination with barred windows. If this is the case, the simplest course may be to breach a wall to get into the apartment. The hallway wall should be opened quickly, and the door left in place, if it is

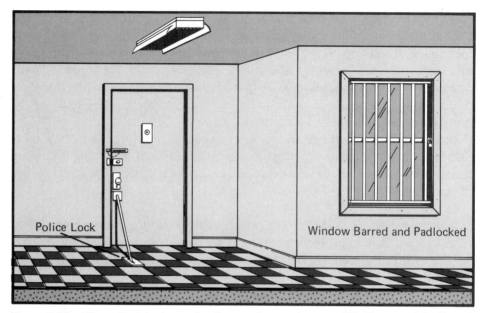

Figure 7.12. Where floor locks are in place, entry can be very difficult, especially if steel doors are used. Entry might be accomplished more rapidly by breaching the wall from the corridor or from an adjoining unit if it can be entered quickly.

known or suspected that occupants must be rescued from the apartment. If an adjoining apartment has been opened or can be entered quickly, it might be easier to breach the wall between the two units. Common apartment walls are usually of lighter construction than corridor walls.

Balcony doors. In many modern apartment buildings, sliding glass doors lead from individual apartments to balconies. When located properly relative to the fire situation, these doors can be used for entry and fire attack. The doors might be equipped with cylinder locks or with some bolting arrangement holding them at the top and bottom. The framing around the doors is usually aluminum or light steel.

A cylinder lock can be pulled or forced as previously discussed. The locations of bolts can be determined by prying with a halligan or claw tool or an axe. Once located, the bolts should be forced with the available tools. Usually, this will spring the doors away from the light framing.

If a door is particularly tough to force, use a flathead axe to drive a pry tool between the door and the framing. Two balcony doors locked to each other also can be opened by driving a pry tool between them or by using a porta-power or rabbit tool (Figure 7.13).

Because there is so much glass involved, care should be taken to avoid straining the glass enough to break it. As a rule, the glass should be broken out for entry into the apartment only when there is a need for immediate rescue, or when the glass is already stained or damaged by heat or smoke. However, when a bar or rod holds the sliding section of a door in place, glass will have to be knocked out to reach and unlock or remove the bar.

Office Buildings

Generally, forcible entry into the units of an office building presents the same problems as entering the units of an apartment house. The age of the

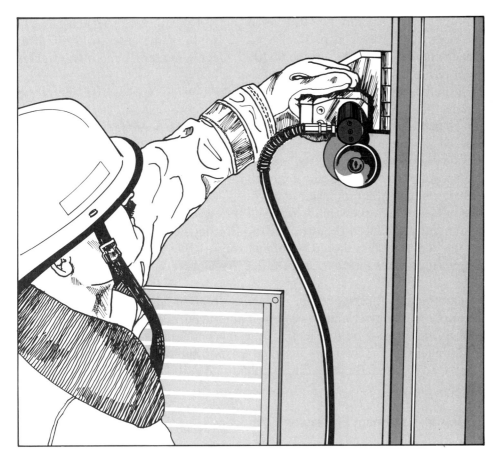

Figure 7.13 A rabbit tool—a small hydraulic spreader—is useful in opening doors.

structure usually determines the type of inside office door, unless the building has been remodeled extensively. In this case, it is important that all fire companies in the immediate area know that such work has been done.

Most office buildings are open to the street during the day but locked after business hours and on weekends. Security and maintenance people could be working in a building while it is closed, but they might not be near the entrances when fire companies arrive. Fires in offices sometimes are reported first by occupants of nearby buildings, so fire companies arrive before people in the fire building are aware of the situation.

The type of door and the material around it determine how entrance will be made. Outside entrances to office buildings are usually similar to those found in stores of the same general age. Modern buildings have tempered-glass doors, and the older buildings use light metal or wooden doors. The doors are forced as discussed under "Commercial Occupancies" in this chapter.

Other Occupancies

Warehouses and factories. These structures usually have roll-up doors at loading platforms and heavy wooden or steel pedestrian doors. In addition, the windows on lower floors may be barred. These types of doors have already been discussed, as have window bars which can be pried apart with hydraulic spreaders or hand pry tools.

Many warehouses and factories are surrounded by fences — usually of the chain-link type. After working hours, fire fighters might have to force their way through the fence before they can get to the fire building. This may only require forcing a padlock or two with a halligan or claw tool, so that a gate can be opened. However, if padlocks on the inside of the gate are difficult to reach, the fence can be bridged with ladders (Chapter 9) and a fire fighter sent inside to force the locks.

Usually, one gate is padlocked from the outside. Its location should be known through prefire inspection. First alarm response routes should be set up so that fire companies arrive at that gate. Where possible, fire companies should arrange to have on-duty plant security personnel open the gates and accompany fire fighters to the involved building in the event of a fire.

A word of caution: in some areas, these occupancies are protected at night by guard dogs roaming inside the fence. The fire department should know, through prefire inspection, where guard dogs are being used; in some communities, ordinances require that the fire department be notified of such protection. A system should be set up to ensure that the agency supplying the dogs is notified as soon as possible after first alarm companies are dispatched.

Combination occupancies. So-called loft buildings, with several types of occupancies, almost always present a double entry problem. After business hours, entry must be forced first into the building itself and then into the individual units. The doors installed in such buildings are similar to those already discussed; those leading to individual units can be heavily barred and bolted.

Sidewalk Basement Entrances

Doors leading to a basement, either from outside a building or from inside on the ground floor, are usually constructed like those already discussed. The one exception is the sidewalk entrance, consisting of two steel doors installed flat, or nearly so, on the sidewalk. The doors open upward to provide a large opening into the basement which can be used to advantage in many fire fighting situations.

Sidewalk basement doors are either manual or automatic. The manual type can open onto a stairway leading down, or can only provide an opening to the basement. The automatic type is rigged to operate with an electrically driven elevator; the doors open as the elevator rises to the street level. A bell and a sign are mounted on the building near every set of automatic sidewalk doors; the sign indicates that the doors are automatic, and the bell rings when the doors are opening.

Sidewalk doors are sometimes difficult to open, especially the automatic type. They are rarely padlocked from above, since the padlock would interfere with pedestrian traffic, but instead are usually locked from below with a bolt or sliding latch. However, if a padlock should be found on a set of sidewalk doors, it can be quickly forced with a halligan or claw tool.

Most often, the forcible entry operation is more difficult. The location of the lock (on the underside of the doors) will be indicated by several bolts or rivets somewhere near the overlapping part of the doors. The doors should be pried apart as much as possible at that location so the lock can be seen. If the lock is of the bolt type, a tool should be placed against it and driven in with a flathead axe. This will tear the lock loose, but it may be a tough job. If the lock is of the latch type that swings parallel to the doors, it might be possible to get a tool onto the swing latch to drive it out of its keeper.

When the lock cannot be forced, the hinges of one door should be attacked. Once these hinges are broken loose, both doors can be lifted off the opening as a single unit. It may be possible to drive the hinges loose with a maul, or to break them off with a halligan, kelly or claw tool driven with a maul or a flathead axe.

If necessary, a power saw can be used to cut through the door to the lock. Cuts should be made around the lock, so the door sheeting can be pried up and the bolt or latch released. Knowing the type of lock involved — through prefire inspection — increases the effectiveness of the forcible-entry operation.

SUMMARY

Prefire inspection and planning are important parts of forcible-entry operations. Truck companies must be familiar with the different types of entrances to buildings in their territory, and with the tools needed to force these entrances. Fire fighters also should be aware of buildings that would present especially difficult entry problems if they became involved with or exposed to fire. Truck companies need training to cope with these problems, using specially designed forcible-entry tools if necessary.

Depending on the fire situation, it might be easiest and fastest to force entry to a building through unbarred windows. Otherwise, doors must be forced, or walls must be breached. The most difficult doors to force are those made of tempered glass or steel; it is often quicker to break through a wall than to open a door constructed of these materials. Thus, it is important that the apparatus carry a full range of tools — from standard hand tools to power saws, hydraulic- and air-powered spreading tools, and (if warranted by possible entry problems) explosive tools.

AERIAL OPERATIONS 8

Each of the three general types of aerial units — aerial ladder, extending platform, and articulated (jointed) platform or "snorkel" — has its own advantages and disadvantages. The type that is best for a particular truck company depends on the operations for which it will be used most often and on the makeup of the company territory. The aerial ladder and extending platform give greater angular reach for a given length, but all three types are versatile and can be used in a number of fireground operations. These include.

- Rescue — removing occupants from windows and roofs; lowering the injured to the ground on litters; placing truck crews on upper floors for search.
- Ventilation — placing crews on the roof and upper floors; knocking out windows with the aerial unit itself or with elevated streams.
- Attack lines — providing access to the building for fire fighters with lines; hoisting lines up to crews in the building; positioning a line for use as a portable standpipe.
- General access — augmenting or replacing stairways and fire escapes for entry to the fire building.
- Hoisting — use as a derrick to hoist sections of hose, standpipe bags, tools, fans, appliances and other equipment to fire fighters on upper floors.
- Elevated streams — directing heavy ladder or platform-pipe streams for initial attack; general fire fighting and exposure coverage; raising handlines for similar operations.

The first three groups of operations are discussed in detail in this chapter; elevated streams are discussed in Chapter 11. The uses of aerial units for general access and for hoisting are self-explanatory. For access, they are especially useful in relieving the load on fire escapes, or when fire escapes and stairways are crowded with evacuating occupants or have been damaged by fire. Hoisting equipment with an aerial unit is quicker than carrying it up several flights of stairs or up the ladder, usually requires fewer personnel and is certainly less tiring.

SAFE PROCEDURE

The next several pages deal with, among other things, the way in which the turntable should be spotted (positioned) for various operations and fire conditions. The positions and aerial unit movements discussed are recommendations, not rules, and must be modified on the basis of the manufacturer's specifications concerning the maximum reach of the unit in relation to the angle at which it is raised. In all cases, the operator must follow the practices established by the manufacturer of the particular unit. Positioning is also affected by the presence of electric wires or trees and the type of surface on which the apparatus is positioned.

The apparatus must be completely stabilized before the unit is raised from its nested position. Generally, this includes locking the brakes, chocking the wheels, and setting the jacks. Again, the manufacturer's recommendations must be followed, and all the steps in the stabilizing procedure must be completed.

The operator must be extremely careful when raising, extending or moving the aerial unit near electric wires. The unit should be watched constantly, to see that it does not touch the wires (Figure 8.1). If necessary, an officer or crew member should be positioned where both the unit and the wires can be seen in order to be able to alert the operator to an impending problem.

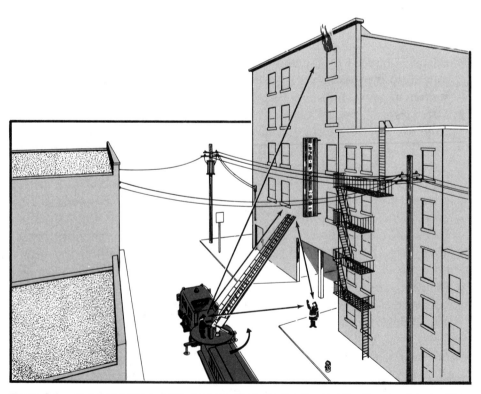

Figure 8.1. Aerial operators must use care when moving the aerial unit near wires or other overhead obstructions.

Figure 8.2. Operators of articulated platforms must watch out for wires and other obstructions when operating at low levels.

An articulating platform presents an additional problem: the elbow of the boom must not make contact with wires on the opposite side of the street from the fire building and from the basket (Figure 8.2). The operator will most likely be watching the basket and might not see the elbow. This problem is most severe on a narrow street when the platform is not being raised very high and the reverse overhang is quite long.

Fire fighters who are not actually on the apparatus must be trained not to touch the aerial unit while it is working. Anyone touching both the ground and the aerial unit can be severely injured or killed by electric shock if the aerial unit makes contact with a power line. Those on the unit are usually safe from shock. Occasionally, though, fire fighters in a basket or on a ladder have been injured when the unit touched an electric wire.

Officers should be aware of areas in which electric wires may present a problem. If necessary, operational training sessions should be held in such areas to determine the safest and most effective ways to raise the aerial unit.

An aerial unit should not be overloaded with personnel or equipment. Here, again, the manufacturer's recommendations must be carefully followed. In addition, officers and crew should be aware of the effect of ladder-pipe or platform-pipe operation on the allowable loading of the unit. Personnel should leave the unit before the pipe is charged, if this is necessary to prevent overloading.

RESCUE

Rescue operations, if necessary, will begin as soon as fire companies arrive on the fireground. Although it is best to remove people by way of interior stairways, there are situations in which the aerial unit must be used to get people out of a fire building. Occupants might be at the windows calling for help or apparently ready to jump when truck companies arrive, or conditions within the building might make it necessary for search crews to evacuate victims through upper-story windows, rather than through interior stairways.

Whatever the reason, the aerial unit should be used for rescue when people are within its reach. If possible in such cases, the apparatus should be positioned for rescue on arrival.

Spotting the Turntable

In Chapter 2, positioning the truck in relation to other apparatus was discussed in general terms. The goal is to get the turntable into a position that will allow the aerial unit to be used most effectively. For rescue, the best position depends on the number and locations of victims and on the wind direction.

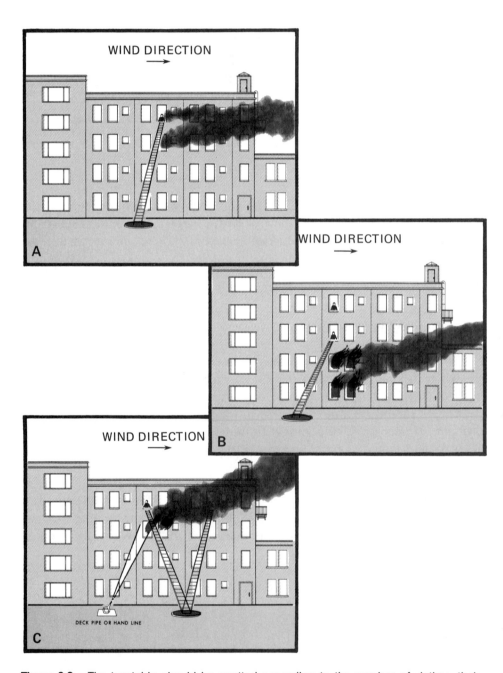

Figure 8.3. The turntable should be spotted according to the number of victims, their location, and wind condition.

When the victims are at a single window, or at several windows that are close together, the turntable should be spotted close to the victims (Figure 8.3A). If the wind is blowing across the front of the building, the turntable should be located upwind from the victims. The smoke from the fire will then be carried away from the approach of the aerial unit as it is raised toward the victims. Both the victims and the operator will be able to see the unit, and the victims will be removed into a smoke-free area. If the fire is upwind of the victims, the turntable must be spotted in the best possible position to get to the victims quickly.

The upwind turntable position is all the more important when fire is issuing from windows below the victims (Figure 8.3B). With the turntable in this position, convected heat and embers will be blown away from the aerial unit and the people using it. If a hoseline is available, it should be directed at the fire, in an attempt to knock it down. Even as a temporary measure, this will protect the aerial unit and the victims until they are rescued. Use of a hoseline also might be necessary when the fire is upwind of the victims.

When the victims are located at some distance from each other, the turntable should be spotted between them (Figure 8.3C). That is, it should be approximately centered between the victims that are furthest from each other. This means that the unit will be operated in smoke part of the time, and the smoke will probably obscure the operator's view of some victims. However, it is still the best position under the circumstances. It would take too much time to spot the turntable upwind to remove some victims, and then respot it to remove the rest. In addition, the arrival and positioning of other apparatus might make respotting impossible.

Sometimes, an aerial ladder can be positioned so that victims can be removed from two or more floors at once. In such situations, the turntable is spotted so that the ladder can be raised parallel to the side of the building at an angle that will provide access to windows on more than one floor.

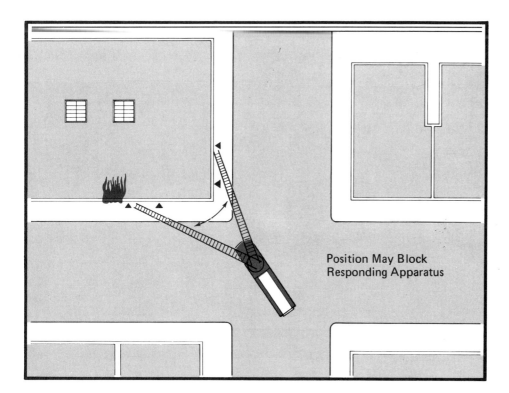

Figure 8.4. When possible, position the aerial unit to cover more than one side of the building.

In some cases, the aerial is positioned at the corner of a building to permit coverage of two sides — for example, the front and one side — at the same time (Figure 8.4).

Raising the Aerial Unit

Once the apparatus is spotted and stabilized, the aerial unit is raised toward the victims who are in the most danger. Normally, this means the people trapped closest to the fire. However, in some cases, occupants who are trapped at some distance from the fire might be in the most danger from heat, smoke and gases. They could be in the path of combustion products being carried by the wind, or these products may be flowing into their apartments or offices through interior shafts and stairways. In any case, the rescue situation must be sized up quickly and carefully and the victims in greatest danger removed first.

One other consideration affects the way in which the aerial unit is raised. If the unit is extended toward occupants trapped at a particular window, then victims at higher-story windows might attempt to jump down to the basket or ladder. This is especially likely when the people trapped above the extended aerial are panicky or when they are only one floor above the extended unit. Victims who jump for a ladder or platform could kill or injure themselves, fire fighters, and other victims, and could put the aerial unit out of service.

An attempt should be made to establish visual or verbal contact with occupants who must await rescue while others are being removed. If they know that they are seen and will be rescued, they might calm down and stay in the building. However, smoke can obscure occupants' view of fire fighters,

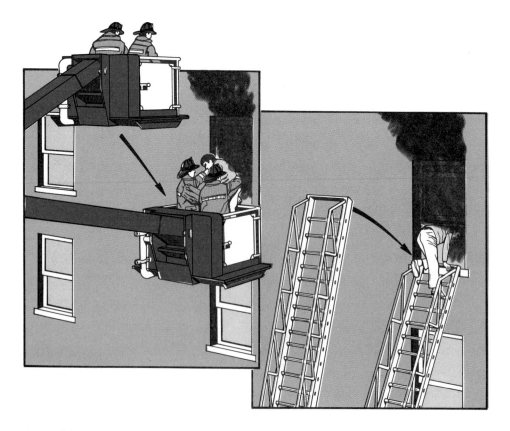

Figure 8.5. When used for rescue, the aerial unit should be kept above the victims until the last seconds of the approach.

and the noise of the fire and fire fighting operations might cover even the sound of battery-powered megaphones. Thus, the trapped occupants in the most peril are those endangered by their own mental state. Again, the situation should dictate which victims must be removed first.

In a normal approach to a window, fire escape or balcony, an aerial ladder is raised and rotated to the building. The elevation is adjusted so the tip of the ladder is aimed at or just above the sill or railing, then the unit is extended to the building. An aerial platform is normally extended so it finally moves horizontally toward the window sill or balcony railing, or up to it from below. These approaches cannot be used for rescue because even those victims who are about to be removed from the building might attempt to jump down or across to the aerial unit before it is in place.

When distance and height factors permit, the aerial unit must first be raised well above the victims. Then, the basket or ladder tip should be dropped down to the victims in the final approach (Figure 8.5). The ladder or platform will not be approaching from below or be level with the victims until the last few seconds before they can reach it.

Placing the Ladder or Platform

During the final approach to the victims, the tip of the ladder or the top rail of the basket should be placed carefully with respect to windows, fire escapes and balconies. The placement should allow trapped occupants to climb onto the unit with maximum ease and safety. It should also ease the job of fire fighters who are assisting or carrying victims onto the ladder or basket.

Ladders. If the ladder tip is extended up above a window sill, or above a balcony or fire escape railing, it will get in the way of victims and fire fighters who are assisting them. Victims would have to climb up onto the window sill or railing and swing over the trusses to get onto the ladder. Further, a ladder that

Figure 8.6. When used for rescue, the tip of the aerial ladder should be placed at or just below the level of the window sill.

is extended into a window might block too much of the window opening, even if it is placed to one side of the opening (Figure 8.6A).

The tip of an aerial ladder should be placed so that it is just at or slightly above the level of the window sill or railing. This allows victims to climb over the sill directly onto the ladder (Figure 8.6B) and allows fire fighters to pass a victim over the sill or railing quickly without having to maneuver the victim up and around the trusses.

When occupants are trapped on a fire escape or balcony directly over the fire, the situation is more acute. In such a case, the aerial ladder should be raised on the side of the fire escape or balcony that is least exposed to fire. In the final approach, the tip of the ladder should be placed against the wall 2 to 3 feet *above* the railing (Figure 8.7A). This will allow victims to use the trusses as hand holds while they climb onto the ladder.

Platforms. The top rail of the basket should be placed about even with or slightly below a window sill or a fire escape or balcony railing (Figure 8.7B). Again, this will allow easiest and safest access to victims and to fire fighters assisting them. When a balcony or fire escape is above the fire, the platform should be raised on the least exposed side, but the top rail of the basket should still be placed about even with the railing.

Imperfectly Spotted Turntable

It might be impossible to spot the turntable for a good final approach — especially to the side of a balcony or fire escape. Nevertheless, it should be spotted to allow as much as possible of the ladder or basket to make contact with the railing. This should also be attempted when the aerial unit will have to be moved, such as from one balcony to another.

Ladders. When only one beam of an aerial ladder will make contact with a window sill, fire escape or balcony, the tip of the beam should be placed above and about 6 inches away from the balcony railing during final approach.

Figure 8.7A. When removing victims from a fire escape or balcony, the aerial ladder should be placed against the wall of the building above the railing.

Figure 8.7B. When using an aerial platform to remove victims from a fire escape or balcony, the top rail of the basket should be placed even with or slightly below the railing.

This will allow the beam to settle on the railing as victims climb onto the ladder. If the beam is placed on the railing, the weight of the victims will cause the unsupported truss to twist downward. With both beams essentially unsupported, the fly section will bend evenly.

Platforms. When only one corner of a basket will contact a balcony or fire escape, the top rail of the basket should be placed a bit higher than the balcony railing. Victims will then be able to get a good hold on the basket as they climb in. The operator should always assist them.

Removing Trapped People

When a number of trapped people must be removed by aerial ladder or basket, they might try to force their way onto the unit. Fire fighters must guard against overloading, which can cause injury and death as well as apparatus malfunction. In many cases, fire attack and venting operations (which should be initiated as soon as possible) will improve the situation quickly. Fire fighters often can calm victims and get them to maintain their positions by pointing out the improvement. If the situation does not improve quickly, previously calm victims can become panicky; then it will be all the more important that aerial rescue operations be performed quickly and effectively. One way to do so is to get more truck companies to the scene as soon as possible. Again, it is essential that first alarm response be increased in high life-hazard areas, so that sufficient personnel and apparatus are available when needed.

Ladders. If possible, at least one fire fighter should be assigned to assist trapped people onto an aerial ladder being used for rescue. The fire fighter should be up with the victims, especially if they are grouped at one fire escape, balcony or window, and should try to space them out on the ladder to distribute their weight. Conscious, able adults should be allowed to climb down the ladder, directed and encouraged by the fire fighter. If available, one fire fighter should precede them down the ladder to the ground.

Small children and anyone unconscious or injured must be carried down by fire fighters. For maximum safety, a backup fire fighter should precede the one carrying the victim — even if the victim is a child (Figure 8.8). The backup person can help the burdened fire fighter maintain balance, can hold the fire fighter against the ladder, can support some of the victim's weight, and, if necessary, can direct the fire fighter in placing his feet on the rungs of the ladder.

When trapped occupants must be removed from two or more locations, the ladder should be kept at the first location until all the victims there have climbed down. Then the ladder can be moved to the next location. An aerial ladder should not normally be moved while anyone is climbing on it.

In some circumstances, however, the people at the second location may be threatening to jump. Sometimes, they will calm down when they see fire companies carrying out rescue work. At other times, the fact that the aerials do not move directly to their location will increase their panic. Also, they may actually be in great danger. In these latter cases, the ladder might have to be moved with people on it.

Such movement should be limited to rotation of the ladder (and slight elevation to clear the sill). It should be accomplished by hand-crank operation unless the hydraulic rotation is very smooth and exact. People on the ladder should be warned that the ladder is about to be turned.

Figure 8.8. Whenever possible, a fire fighter carrying a victim down an aerial ladder should have a backup for assistance as needed.

If rotating the ladder will place it within safe reach of others attempting to get out of the building, it should be positioned for them. If not, the ladder should be kept out of their reach so as not to tempt them to jump for it. The act of turning the ladder toward other victims might have a calming effect on them, for it could be their first indication that they have been seen. Once the people who have been removed from the building are clear of the ladder, it can be extended as necessary to reach others.

If the fire situation is bad, fire fighters might not be able to climb up the ladder in time to keep victims from overloading it. When the loading might exceed the manufacturer's recommendations, the aerial ladder should be supported from below with ground ladders. At least one support should be placed at the end of the bed section; additional supports may be required at the ends of other sections if the overload is severe. The aerial ladder might also have to be supported when it makes a small angle with the ground.

Platforms. At least one fire fighter should be assigned to make sure trapped occupants do not overload the aerial platform by climbing into the basket all at once. The situation can be difficult if the loaded basket must make a trip down while some victims remain in the building. They will all want to go in the first group and must be controlled. Often, if a fire fighter stays with them, they will keep calm until the basket returns.

When some people must remain in the building while the basket makes a trip down, the operator should move the basket upward first and then away from the building to discourage victims from jumping at it. The basket should be returned to the waiting occupants as soon as possible.

A fire fighter with an unconscious victim should hold the victim in the basket until they reach the ground. Young children should be held or carried by fire fighters in the basket until the descent is completed. This is especially important when conscious adults are to be lowered at the same time. A child or an unconscious adult lying in the basket could be trampled by others in their rush to get out of the building.

In severe situations, several unconscious victims might have to be removed from the building at the same time, and it may not be possible for fire fighters to accompany them. Unconscious victims should be sent down before conscious victims. They should be placed in the basket carefully and taken to the ground quickly to receive medical attention. However, fire fighters in the building must be careful not to overload the basket. If necessary, the unconscious victims should be sent down in groups.

Removing Victims by Litter

The fire situation may allow injured or unconscious victims to be removed by litter. It is best to carry a litter down an interior stairway if the stairway is wide enough and the victim is found on a lower floor. However, when the litter must be carried down many flights or when the stairways are narrow and twisting, the victim can be brought to the ground with the aerial unit. A stokes litter (also called a stokes basket or stretcher) is preferred for such operations.

Ladders. To raise the litter, the ladder is first extended two or three rungs; then a lifeline is passed under the bottom rung of the bed section and brought out immediately over the remaining rungs (Figure 8.9). The line is then passed up over the ladder and dropped down between the first and second rungs of the fly section. A hose roller placed over the rung will help prevent wear and tear on the rope. (Some aerial ladders have a special double rung and guide for this purpose.)

The line is tied to the sling on the litter under the ladder. As the ladder is raised, rotated, and extended up to the building, the rope is fed up the ladder to hold the litter in position. The weight of blankets and first aid equipment in the litter will help hold it in position. A small-diameter guideline attached to the litter will help prevent spinning.

The ladder is raised and adjusted until fire fighters in the building can get hold of the litter and pull it in through a window or onto the roof, fire escape, or balcony from which they are operating. Slack in the lifeline should be passed to them as they pull in the litter. Once the litter is in the building, the lifeline should be doubled, tied securely to the bottom rung of the bed section, and monitored by a fire fighter.

When the litter is ready to be lowered, it is only necessary to pull the aerial ladder away from the building or to extend it slightly to take up the slack as fire fighters in the building guide the litter out. Then the aerial is rotated away from the building and the ladder retracted until the litter contacts the ground.

This operation does not require block and tackle, so there is no time lost rigging it to the ladder and lowering by hand. With a little practice, the movements can be performed very quickly.

Platforms. The litter can be raised in the basket. The litter should then be taken into the building, and the victim should be firmly strapped into it. The litter should then be laid across and tied to the basket with rope hose tools,

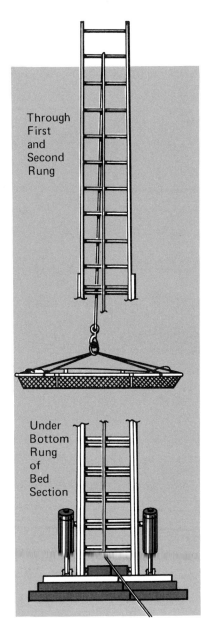

Figure 8.9.

Figure 8.10. Aerial platforms may be used to remove victims in litters. The litter should be tied in and held by a fire fighter.

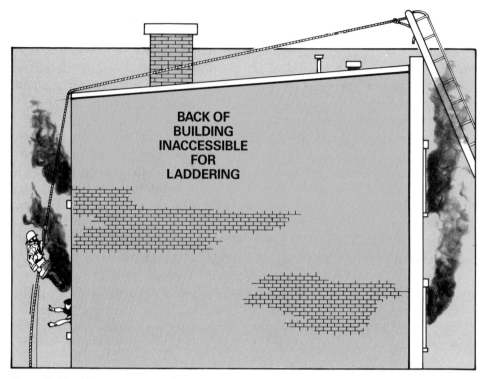

Figure 8.11. When necessary for rescue, an aerial ladder or platform can be used as the anchor for a lifeline.

hose straps, or rope (Figure 8.10). If injuries require that the victim's head be kept raised, the litter can be placed vertically in the basket and then tied in. In either case, a fire fighter should descend with the litter and hold it as the descent is made.

Lifeline Anchor

One rare but sometimes necessary use of the aerial unit is as a lifeline anchor. The aerial unit might be used in this way when one or more victims are trapped at the rear of a building, the aerial unit cannot be positioned to reach them, and they are higher than the reach of the longest ground ladders. If nothing on the roof will serve as a lifeline anchor, an extended aerial can be used.

With an aerial ladder, the lifeline should be tied to the rung closest to the point of contact with the roof, to prevent bending of the ladder (Figure 8.11). With an aerial platform, the line should be secured to a strong member of the basket or boom assembly. It should not be tied to the top hand rail.

The lifeline is taken across the roof to the far side and dropped down. The rescuer can slide down the roof rope or be lowered by others to get to the victim, depending on the number of fire fighters present and on department policy and training. The rescuer should wear a life belt and carry one for the victim. Once the rescuer has slid or been lowered down to the victim, the extra belt is put on the victim and hooked into the rescuer's belt hook, and both descend. If the victim should get loose or become unconscious, the fire fighter's life belt will keep the victim from falling to the ground.

Sometimes, an excited victim will grasp the approaching fire fighter in a very tight grip with arms and legs and the rescuer will be unable to use the extra belt. In this case, the descent should be made as quickly as possible.

Aerial units can be used in many ways to rescue victims from buildings. Efficient fireground operation will be promoted by prefire on-site visits, during which the turntable is spotted in different positions, and the movements which might be required are practiced.

VENTILATION

In most fire situations, the ventilation of a multistory building begins at the roof. An aerial unit can be used to place fire fighters on the roof, but, as noted in Chapter 5, other means should be used if they are available. The aerial unit

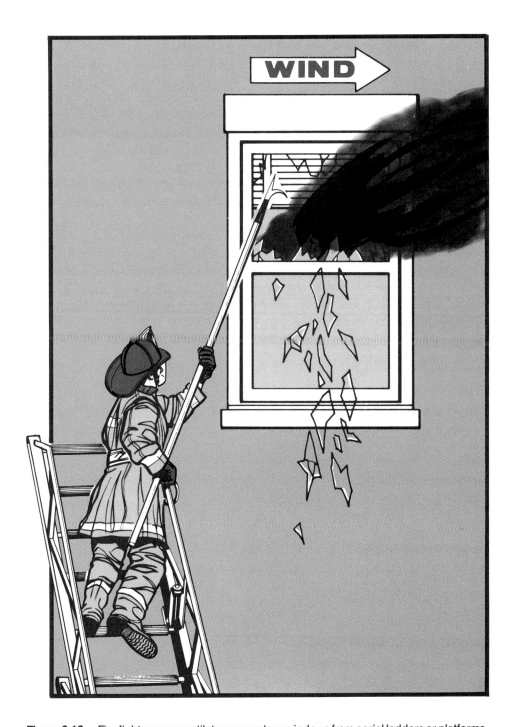

Figure 8.12. Fire fighters can ventilate upper-story windows from aerial ladders or platforms.

should be reserved for possible rescue duty, at least at the beginning of fire fighting operations.

With roof venting initiated or completed (or, in some cases, deemed unsafe), the windows on the top floor are opened or removed to ventilate that floor. If truck personnel cannot get to those windows, the aerial units not being used for rescue can be used for top-floor venting. Windows on the top floor (and lower floors, if necessary) can be knocked out by a fire fighter on the aerial unit, by the ladder or platform pipe stream, or with an aerial ladder.

Removing Windows

One fire fighter on an aerial ladder or in a basket can quickly knock out a row of windows with a 6- or 8-foot pike pole. The fire fighter should be secured to the ladder or to the basket side rail with a life belt (Figure 8.12). A leg lock should not be used when operating from a ladder; if used, any unintentional extension or retraction of the fly section could cause serious injury. In a basket, the life belt will prevent injury if the aerial unit moves unexpectedly while the fire fighter is reaching out toward a window. More then one fire fighter has been thrown out of the basket in such a situation.

Use of a pike pole allows the fire fighter on the basket or ladder to operate several feet from the window to be opened. The fire fighter should be placed to one side of the window, about even with the sill. If wind is a factor, placement should be on the windward side, for protection from heat, smoke, gases and falling glass.

When more than one window must be opened and the wind is blowing across the face of the building, the first window opened should be the one furthest downwind. Moving from window to window into the wind allows the combustion products leaving the windows to be carried away from the fire fighter, whose vision will not be obscured as it would be if upwind windows were opened first.

Venting with Streams

In some situations, initial size-up indicates that it would be extremely dangerous to place a venting crew on the roof of the fire building. One such situation would be in a building heavily charged with smoke but showing no flames — that is, at a suspected smoldering fire. A backdraft could trap crews on the roof or injure a fire fighter venting with a pike pole from an aerial unit. In these cases, the windows on the top floor of the building can be raked out with hose streams directed from the aerial unit.

A ladder or platform pipe or a handline can be used to knock out the windows. Solid streams should be used for this operation. The ladder tip or basket should be placed away from the building so that a fire fighter on the aerial unit will not be endangered by flying glass or possible backdraft. As before, the furthest leeward (downwind) window should be knocked out first if a wind is blowing across the face of the building.

Venting with an Aerial Ladder

The aerial ladder itself may be used to knock out windows in a fire building. One fire fighter — the operator — can provide much ventilation in a short time by properly extending the tip of the fly section through each window in turn, breaking out the glass, and then moving the ladder to the next window.

Figure 8.13. When using aerial ladders to vent windows, the window furthest downwind on the top floor should be knocked out first.

Position and sequence. If it is known upon arrival that the aerial ladder will be used for venting, the turntable can be spotted for maximum effectiveness. When the wind is blowing across the face of the building and exposures are located close to its downwind side, the turntable should be spotted just upwind of the closest exposure (Figure 8.13). This allows the aerial ladder to be used on the fire building and (if necessary) on the exposed building. However, if the fire building is relatively wide, the apparatus should be positioned closer to the center of the building.

If the fire building is very wide, or if there are no downwind exposures, the turntable can be spotted at the center of the building to allow maximum reach and to put it within range of most top-floor windows.

As in other venting operations, the window furthest downwind should be opened first, and the ladder should then be worked back into the wind. This will allow maximum visibility for the ladder operator and allow combustion products to be carried away from the ladder. However, if fire issues through a window, the operator must move the ladder away from it quickly.

If the next lowest floor is also to be opened with the aerial ladder, the top floor should first be opened completely. Then the ladder should be moved to the furthest downwind window on the floor below and the procedure repeated.

Knocking out windows. The first step in knocking out a double-hung window is to extend the ladder tip through the upper section. If possible, the ladder should be extended in far enough to push away obstructions such as curtains, blinds and shades (Figure 8.14). If these obstructions are not removed, they can limit the venting action. After the tip has been extended into the window's upper section, the ladder should be lowered to break through the window frame and the glass in the lower section. Then the ladder should be retracted and moved to the next window.

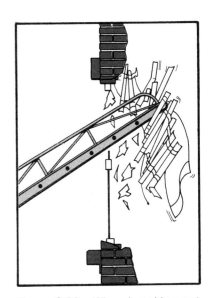

Figure 8.14. When knocking out windows with aerial ladders, be sure to remove obstructions to the window opening.

In some modern structures, large picture windows are installed between pairs of small casement windows. The ladder operator should ignore the smaller windows and remove the picture window completely. The ladder tip should be extended through the top center of the window and then lowered to the sill to clean out most of the glass. If the picture window is very wide, the ladder can then be moved from side to side to clean out any remaining glass.

The operator must constantly observe the tip of the ladder to make sure of proper penetration into the window, and must be able to retract the ladder promptly if it becomes engulfed in flame. The operator must be extremely careful not to damage the ladder while using it for venting. Although speed is essential, careless operation, excessively fast movements, and poor judgment of angles can cause serious damage to the ladder.

If the ladder is extended too far into a window, it might jam into the ceiling and get stuck. The ceiling will then have to be cut away to free the ladder; by the time this can be accomplished, the ladder may very well be damaged by flames and heat.

The ladder can also be damaged if the operator tries to use it to break through a window frame made of steel or some other strong material. When a ladder will not break through a window frame easily, it should be retracted, moved, and extended to break out the panes one by one.

Sometimes, because of the location of either the window or the apparatus, a window must be approached at an acute angle. Then, only the inside truss of the ladder should be used to break through the glass, and the ladder should be moved, extended, and retracted slowly. Otherwise the ladder will bind in place. The same precautions must be observed when narrow windows are being knocked out, no matter what the angle of approach. Ladder tips should never be forced through an opening.

Safety. Large amounts of glass and debris will fall to the ground when a building is successfully vented with an aerial ladder. Shards of glass and chunks of debris can slide down the ladder toward the operator, who must be aware of this possibility. In a strong wind, shards of glass can scale a good distance toward fire fighters performing other operations (Figure 8.15). They should know when a building is to be vented by ladder and attempt to stay clear of the immediate area.

Timing of the venting operation will be especially important if windows over the building entrances must be knocked out. Fire fighters, who most likely will be using the entrances to take lines and tools into the building, should be warned to stay clear when the venting operation is to begin. Crews caught unexpectedly in a shower of glass and debris should keep their heads down with their arms close to their sides. *They must not look up.* They should move as close to the wall as possible, seek protection in doorways and under overhangs, and proceed only when advised that it is safe to do so.

HOSE OPERATIONS

Aerial ladders and platforms can be used to great advantage in getting personnel and hoselines into the upper stories of a building. Place an aerial ladder to allow engine crews to take lines up into the building. An aerial platform can lift fire fighters into the building, after which they can haul the hose up by rope. Either aerial unit can be used to hoist fire fighting lines, standpipe rolls or bags, and allied equipment to crews already in a building, and either unit may be used to place a line as a portable standpipe.

Figure 8.15. When window-venting operations are being carried out, crews on the ground should be warned to stay clear.

Lifting Personnel and Equipment

When engine crews are to climb an aerial ladder carrying a line, the tip of the ladder should be placed even with the sill of the window they will be entering. This placement is similar to that for rescue operations, but in this case the ladder should be placed as close as possible to one side of the window to allow fire fighters carrying hose and wearing air masks to get in through the bottom of an open double-hung window.

If the ladder is placed so that two or three rungs extend up over the sill, part of the window opening is blocked off. Fire fighters will have difficulty entering, and their air tanks could get hung up on the upper part of the window. The entire window might have to be removed to provide sufficient access, which means that crews will have to enter the building through an area strewn with broken glass.

Even when the aerial ladder is properly placed, the window opening might not be large enough to allow easy entry. Then, in spite of the broken glass, the window should be removed. It is most important that fire fighters be able to get off the ladder and into the building without mishap.

One section of hose can be tied to the ladder before it is raised, to eliminate the need to carry the hose up to the window. For this, the first section (50 feet) of hose is tied to the ladder with a rope hose tool. The nozzle, the center of the hose, and the first coupling should be tied to the first rung so that they will be immediately below the window sill when the ladder is raised. The hose will hang below in two loops. The ladder is then raised to the window, and fire fighters climb up and in. They can then get hold of the nozzle, loosen the rope hose tool, and pull in the first section. If additional hose will be needed, the engine crew should carry a rope and a hose roller with which to hoist it up.

The aerial ladder can be used to lift sections of hose to fire fighters already in the building. The hose should be tied to the ladder as described above and the ladder raised higher than the fire fighters so the hanging loops of hose are easily accessible. Once the crews have hold of the loops, the ladder should be lowered to them so they can pull in the loops and release the rope hose tool.

Similar operations can be performed with an aerial platform. Engine crews can be carried up in the basket with sections of hose tied to the front of the basket. Hose and appliances can be lifted in the basket to crews already in the building. The aerial lift by ladder or platform is usually much faster than climbing and carrying equipment up stairways or ladders, especially above the third or fourth floor.

Using Hose as a Portable Standpipe

For this operation, the aerial platform is raised to a window or balcony. Hose that has been carried or lifted into the building is connected into the platform's 2½-inch outlet or master-stream nozzle connection. The hose can then be used by engine crews within the building.

An aerial ladder is used to haul one end of a 2½- or 3-inch hose up to fire fighters in the building. The hose should be tied, at the end and at a point 10 to 15 feet further back, to the top rung of the ladder with a rope hose tool. The ladder should be raised to the window or balcony, where fire fighters can get hold of the loop and loosen the rope hose tool. The hose should be pulled in onto the floor and secured to a window sill with the rope hose tool. The ladder is then free to move away for other operations. Hose that has been carried or lifted into the building can be connected to the hose sent up by the aerial ladder and used for fire fighting operations.

Either of these operations can be carried out with a water thief. This appliance will permit the use of one 2½-inch and two 1½-inch or 1¾-inch lines for fire attack.

Both types of "portable standpipe" will increase the effectiveness of standpipe fire fighting operations when the fire floor is not more than two or three stories above the maximum reach of the aerial unit. Aerial hose operations, in general, promote cooperation and coordination between truck and engine companies. Combined training sessions should be conducted to increase the effectiveness of these operations.

SUMMARY

The aerial unit is an extremely versatile apparatus that can be used in rescue, venting, fire attack, and exposure protection operations. It can be used to provide access to the upper floors of a building and as an equipment lift. It also can be used to remove trapped occupants from a fire building, as well as to carry out some fire fighting duties in situations that would be extremely hazardous to personnel.

However, the aerial unit itself must be used with care. The apparatus should be completely stabilized before the unit is raised. The unit should be kept from contacting trees or electric wires and should not be overloaded. It should always be operated according to the manufacturer's recommendations. Perhaps most important, the truck company should be well trained in working with the aerial unit and aware of what it can and cannot do.

GROUND LADDERS   9

For many years, ground ladders (also called portable or wall ladders) were the mainstay of fire department laddering operations. Today, even with aerial ladders and platforms readily available, ground ladders have an important place in truck operations. Many fire departments use only ground ladders, especially departments located where buildings are not more than three or four stories in height.

The main advantage of a ground ladder is its portability. It can be carried to positions that could not be reached with the aerial apparatus. Where a truck company has both ground ladders and an aerial unit, the ground ladders can be used for lower-story operations, freeing the aerial unit for upper-story work.

Several ground ladders can be carried on the apparatus and raised at two or more positions simultaneously. By contrast, a fixed-turntable aerial unit must be raised from position to position and assigned to fireground operations in sequence, one at a time.

The major disadvantages of ground ladders are their limited reach and the personnel required to raise them. Ground ladder operations start becoming inefficient and time consuming when the ladders must be long enough to reach above three floors. From two to six members of the truck company are needed to raise a ground ladder, as compared with the single fire fighter required to raise an aerial unit.

GROUND LADDER OPERATIONS

Within their height and personnel limitations, ground ladders can be effective in a number of fire fighting operations. Some of their uses are the same as those of aerial units, but others are unique, including:

- Gaining access to the fire building and exposed buildings
- Advancing hoselines when stairways are being used by people escaping the building
- Replacing damaged stairways
- Removing trapped victims
- Removing people from crowded fire escapes
- Getting from one roof level to another
- Bridging fences, narrow walkways, courts and alleys
- Forcing entrance through inward-opening doors, display windows, and their enclosures
- Ventilating, by knocking out store, apartment and office windows; by knocking out the glass or transom below a skylight; and by pushing down ceilings from above
- Transporting the injured, in place of litters
- Reinforcing weakened building features

These uses of ground ladders are discussed, by operation, in the following sections. The rest of this section concerns ground ladder operations in general.

Ladder Work

Ladder work is a truck company responsibility. If a truck company has an aerial apparatus, the ground ladders will be transported on it. However, many departments use only ground ladders — they do not have aerial units — which can be transported to the fireground on pumpers (engine company apparatus). In such cases, ladder work should be assigned to fire fighters who will be performing truck duties.

If laddering duties are not specifically assigned, rescue operations or fire attack operations can be unnecessarily delayed. For example, if too few fire fighters take it upon themselves to raise ladders, search and rescue operations could be delayed while search crews wait for ladders. If too many fire fighters decide to raise ladders, too few attack lines will be advanced to the seat of the fire.

Thus, whether or not a department consists of separate engine and truck companies, their duties — and especially laddering duties — should be assigned to different groups. However, engine and truck personnel should hold combined training sessions because they must work so closely on the fireground. In particular, ladder operations affect the efficiency of most other fireground operations.

Ground ladders are available in a number of types and lengths. Which particular ones are carried by a company will depend on standard recommendations, company experience, and the types of structures that comprise the company's territory. It is a good idea to carry at least two ladders of the most-used length and type.

Handling Ground Ladders

There are several ways to carry a ground ladder from the apparatus to a building. The method used in a particular instance should be the one that requires the least maneuvering and the least time. Except for pole (Bangor) ladders, there should be no need to lay a ladder on the ground after carrying it to a building and before raising it.

One way to avoid extra movement in raising a ground ladder is to carry it from the apparatus with the rungs and trusses parallel to the ground. Once the butt is in position near the building wall, it is lowered to the ground while the upper part of the ladder is raised from shoulder height. The ladder then can be pivoted and extended if necessary. With practice, this raise can be performed in one smooth nonstop motion.

Two-person and three-person ladders also can be quickly placed for the beam raise. The ladder is turned with its rungs parallel to the building wall as it is lowered to the ground. The butt of one beam is secured, and the ladder is raised to the vertical position. This raise is often advantageous when the area above the ladder position is restricted by wires, trees or other obstructions.

In preparation for the raise, a two-person ladder is carried from the apparatus with fire fighters on opposite corners (Figure 9.1A). A three-person ladder is carried with two fire fighters at each end of one truss, and one in the middle

Figure 9.1. Ladders carried in flat position are easily handled and quickly raised.

of the other truss (Figure 9.1B). Four-person and larger ladders are carried by two fire fighters at each end (Figure 9.1C), and additional crew members at or near the middle (Figure 9.1D).

Some ladder manufacturers specify where the fly section of an extension ladder should be (toward the building or away from it) when the ladder is raised. These recommendations should be followed. If the manufacturer makes no recommendation, position the fly section on the outside, away from the building, for maximum safety in handling. This places the halyard between the ladder and the building, so the ladder will move toward the building as the halyard is pulled to extend it (Figure 9.2).

If the ladder should get away from the fire fighters, it will fall against the building and not out onto crews and apparatus. However, this requires that the ladder be extended to the desired length before it is moved toward the

Figure 9.2. Maximum safety is obtained by extending a ladder with the fly section on the outside.

building. Well-practiced members of truck companies can safely raise extension ladders with the fly section on the inside, extending the ladder as it is being moved toward the building.

The climbing angle is determined by the height of the raise and the distance from the butt to the building. There are a number of ways to establish the proper butt distance for a good climbing angle. However, most result in the butt-to-building distance being one-quarter of the height of the raise (not one-quarter of the ladder length). Figure 9.3 shows a 50-foot ladder raised 40 feet; base of ladder out 10 feet from building — not 12½ feet.

Safety

Truck crews raising ground ladders should be careful of overhead obstructions, especially electric wires. They should watch the tip of the ladder, not the butt, as they are raising it. (A very small movement of the butt can result in a wide swing at the tip; the longer the ladder, the more pronounced this effect is.)

Once the ladder is raised, it should be tied in to the building or braced by one fire fighter as others climb it. Either action will keep the ladder from moving while being used.

A ground ladder should not be overloaded. At normal climbing angles, fire fighters should be at least 10 feet apart on the ladder. If the climbing angle must be decreased for some reason, the distance between those on the ladder should be increased to 20 feet. Fire fighters carrying hose up a ground ladder should be at least 20 feet apart, at any climbing angle.

Once a ladder has been used to enter a building, it should be left in place as an exit, unless it is required for rescue at a nearby location. It should then be returned to its original position as soon as the rescue has been effected.

RESCUE

Ground ladders are used in several ways to help remove people from a building. The ladders can be placed at windows, fire escapes or balconies so occupants can climb down or be carried down. Ladders also are used to bridge between buildings for access to adjoining roofs, windows or balconies.

No matter how ground ladders are to be used for rescue, the first ladders should be raised to the victims in most danger. As previously noted, these are usually the victims closest to the fire, on or directly above the fire floor. However, smoke and other combustion products may be endangering victims further away from the fire, so that they, too, might require immediate attention. Observation of the fire situation will indicate where the first ground ladders should be raised.

The Normal Raise

It is important that the ladder crew select a ladder of the proper length for the raise — especially when a straight ladder (rather than an extension ladder) is being used. There is little sense in tying up a long ladder for a comparatively short raise. This could strand victims on higher floors who cannot be reached with the remaining (shorter) ladders. In addition, a ladder that is too long for the raise must be placed at an awkward climbing angle or placed so that it blocks part of the window opening.

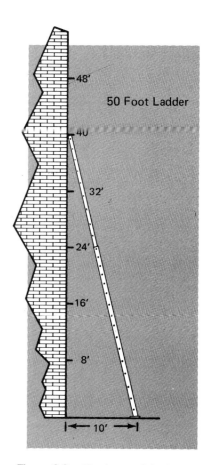

Figure 9.3. The base of the ladder should be out from the building about one-quarter the distance that the ladder is raised.

Training sessions at buildings in the company territory will help in determining which ladders might be needed at different locations. This depends on the building floor heights, which can vary from 8 to 20 feet.

As a ground ladder is raised for rescue, it must be kept out of the reach of people who are to be removed. This is especially necessary when the ladder is being raised past some victims to reach others on the floor above (Figure 9.4). It is almost impossible to place a ladder properly while someone has hold of the upper part. The ladder should be raised in a vertical position away from the building, pivoted, extended if necessary, the tip lowered to the victims, and the butt quickly moved to proper position.

Figure 9.4. When raising a ground ladder for rescue, keep the ladder out from the building so that victims cannot get hold of the ladder and hinder operations.

When a ground ladder is placed at a window for rescue, the tip should be at or just over the level of the sill (Figure 9.5).

In placing a ground ladder at the front railing of a balcony or fire escape, or at the wall beside either of these features, two to four rungs should extend above the railing to provide a good handhold for victims climbing onto the ladder and for fire fighters working with victims. This gives victims a greater sense of security, as do the trusses on an aerial ladder. Ladders raised to a roof should also extend two to four rungs above the roof wall for the same reason.

The Emergency or Hotel Raise

In some situations, a ground ladder is positioned to remove people from several floors at once. Usually, a 40-, 45-, or 50-foot pole (Bangor) ladder is used for this raise, so the raised ladder can be held in place by fire fighters pushing it against the building with the poles (Figure 9.6). However, other types of ladders can be used and held in place with "U" (crotch) poles or, as a last resort, with long pike poles.

In the emergency raise, the ladder is brought to the building, raised, extended, and set against the building with the butt about 4 feet from the wall. Fire fighters on the poles and at the sides of the ladder then hold it against the building while victims climb down. No fire fighter should be positioned between the ladder and the wall.

Victims near the top of the ladder should climb onto and down the outside of the ladder. Those on lower floors can use the inside of the ladder. (It is for this reason that no fire fighter should be under the ladder.) Fire fighters bracing the sides of the ladder should watch the climbing victimns, in case any should fall.

In the emergency raise, it is especially important that the ladder be kept clear of waiting victims as it is being raised. It is best to set the ladder further away from the building than necessary, raise it, lower it against the building, quickly move the butt in toward the wall, and then secure the ladder against the building and at the butt. This will prevent victims from grabbing the ladder before it is in place.

When no eaves or cornices affect the positioning of the ladder, it should be quickly extended to its full length, locked in, and then placed against the building. This way no time will be lost in attempting to extend the ladder to just the right length before it is placed.

The emergency raise can be very effective in quickly moving many people from windows, balconies and fire escapes, provided they are physically able to climb down. With this raise, aged, handicapped and very young victims usually must be carried down other ladders or interior stairways.

Ground Ladders as Exits

If ground ladders will reach to floors being searched for victims, they should be raised as exits for fire fighters with victims and as emergency exits. This should be done even if fire fighters engaged in search have entered the building through interior stairways. The ladders provide additional exits in areas somewhat removed from the stairways and allow fire fighters to get victims to fresh air quickly without having to work their way through the building. The ladders also provide fire fighters and victims with alternative exits in case the stairways become impassable.

Exit ladders should be raised at corridor windows if their locations are known. Fire fighters should look for ladders at these central locations. Alter-

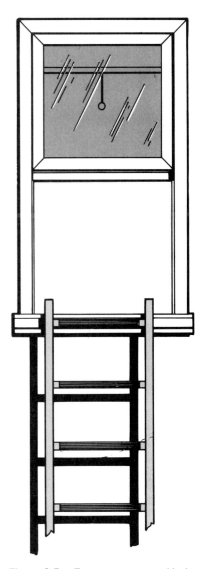

Figure 9.5. For rescue, ground ladders should be placed with the top rung at or just over the sill.

native positions include just about any window on the fire floor and the floor above the fire.

Two-way radio communications are helpful in these operations. Fire fighters inside can use their radios to call for ladders at certain locations. The officer in charge can use the radio to advise those inside that ladders have been placed at certain positions. Where radios are not available, visual and verbal contact, initiated by fire fighters working inside, can be used to properly locate ladders.

Bridging

Ground ladders can be placed horizontally to bridge the space between two buildings or between two sections of a single building, over an alley, walkway, narrow court, or air shaft (Figure 9.7). Once the ladder is in place, trapped occupants can crawl across it to safety. The "bridge" can go from window to window, from roof to roof, or from one balcony to another.

Either straight or extension ladders can be used for bridging. An extension ladder should be long enough to allow sufficient overlap between sections, but not so bulky or heavy that it is difficult to handle — especially if it is to be carried up through a building. Hoisting a ladder by rope to where it will be used is much quicker than trying to maneuver a fairly long ladder up several flights of stairs.

When a ladder is being hoisted with a rope, the rope should be tied several rungs below the top. When the rope is pulled up, these rungs will extend above the window sill or roof top, allowing truck crews to get a good hold on the ladder and use some leverage to get it in.

Bridging techniques. There are several ways to position ladders as bridges. The method used in a particular instance will depend on the features to be bridged and the personnel available.

A straight ladder can be placed from window to window or roof to roof by simply pushing it tip first across the intervening space. Several fire fighters should place their weight on the butt end to keep the tip from dropping below the victims.

An extension ladder can be extended horizontally across the space. Again, several fire fighters should press their weight against the base section while another uses the lanyard to send the extension across to the victims. Since the ladder will be moving toward a group of excited, possibly panicky people, it should be extended quickly and forcefully. Then, even after it is grabbed by the victims, it can be run out to a secure position on the window sill or roof wall. Fire fighters should attempt to make verbal and visual contact with the waiting occupants, to calm them and, if necessary, tell them how to place the ladder tip properly.

As the tip of the ladder nears the victims, truck crews holding down the base or butt should exert extra force in case a victim jumps out onto the ladder. It is sometimes impossible to hold a long ladder with a heavy person at the tip, but the extra force at the base helps maintain the ladder in position. Again, contact with those waiting to be rescued often averts such difficulties.

It may be possible for a fire fighter to get above the ladder, attach a rope to its tip, and by manipulating the rope keep the tip above the level of the sill, railing or roof wall that is to be reached. This will help keep victims off the ladder until it is in place. Once the ladder is properly positioned, the fire fighter should release the rope so that it does not interfere with people getting on the ladder.

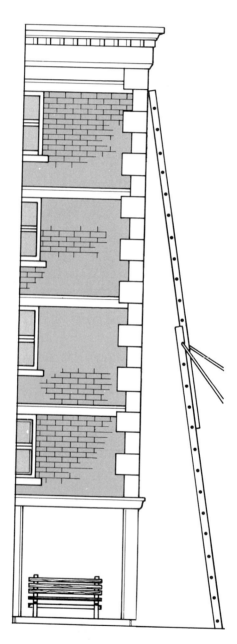

Figure 9.6. The hotel raise of a pole ladder can be useful in removing victims from a building by enabling them to climb either side of a ladder from several floors at a time. No fire fighter should be under or behind the ladder.

It is possible to place an extension ladder from one roof to another by extending it to the proper length and lowering it across to the adjoining roof. This usually requires ropes tied to the tip and another ladder fastened to the butt of the bridging ladder. The second ladder serves as a fulcrum in controlling the descent of the bridging ladder. Keep in mind that this is a time consuming operation, even for a well trained crew.

Balcony to balcony. Straight and extension ladders are also used to bridge the space from one balcony to another across the face of a building (Figure 9.8). If the two balconies are on the same level, they can be bridged by the techniques discussed above. In some instances, it might be possible to bridge from a balcony at one level to a balcony at a higher level. In either case, the ladders should be tied in to railings with rope hose tools or ladder straps before they are used, and during use must be held firmly in place by fire fighters.

Bridging a fence. Although not normally required for rescue operations, there is a bridging operation to get fire fighters over a fence quickly. This operation can be used to gain access to factories and warehouses surrounded by fences with locked gates, to locations in storage yards far from gates, to fenced-in areas found behind many types of structures, and in general to cross any fence at any point.

Two ladders are required. The first ladder is placed against the outside of the fence, and a fire fighter is sent to the top. The second ladder is passed up, butt first, to the fire fighter who lifts it over the fence and places it against the inside. When two different sizes of ladders are used, the heavier one should be placed first, and the lighter one passed up and over the fence. The ladders should be fastened together near the top with a rope hose tool or strap.

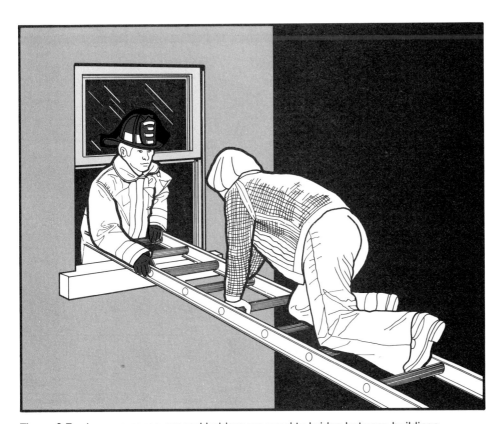

Figure 9.7. In some cases, ground ladders are used to bridge between buildings.

When bridging a metal fence (such as a chain-link fence) at a working fire, truck crews must be wary of electric wires that might have burned through and dropped onto the fence. A severe shock can be transmitted through the fence to working fire fighters, especially if they are using metal ladders. It may be necessary to check some distance along the metal fence, in both directions, because a charge can be transmitted a long way.

Bridging is not associated with the usual ladder operations. Its techniques should be practiced by truck companies to develop proficiency and determine which are best for the normal complement of personnel. Prefire inspection will assist truck companies in determining where such operations might be utilized and which would be the most feasible.

VENTILATION

Ground ladders can be placed to provide venting crews with access to roofs or windows. The ladders themselves can be used to knock out windows and to knock down ceilings.

Using Ladders for Access

When a ground ladder is placed to a roof, two to four rungs should extend up over the roof's edge to provide a good handhold for fire fighters climbing

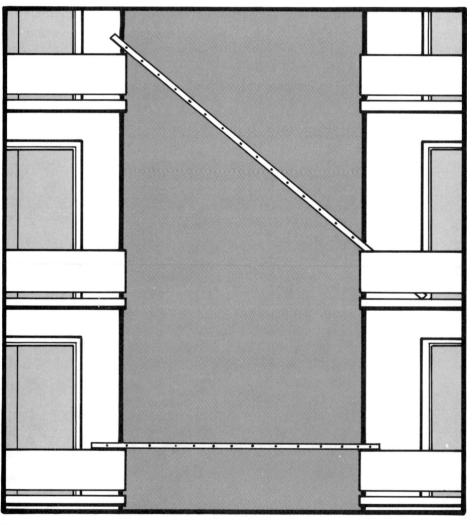

Figure 9.8. Ground ladders can be used to move victims from one balcony to another. Truck crews must hold the ladder securely.

on or off the ladder. For opening or knocking out a window, the ladder should be placed so that the fire fighter on the ladder is upwind from and slightly below the window. This will keep the fire fighter out of the path of flying glass, smoke and gases. When the windows on two opposite sides of a building are to be opened, those on the downwind side should be opened first (see Chapters 4 and 5).

Using Ladders as Venting Tools

Either straight or extension ladders can be used to knock out windows for venting. An extension ladder is extended to the proper length, pulled back, and then pushed through the window. This will drive most of the glass and debris inside and result in a good-sized opening. By moving quickly from window to window, truck crews can ventilate a fairly large area in a short time. Even one fire fighter with a short ladder can rapidly ventilate the upper floor of a two-story dwelling.

Ground ladders also are used to quickly knock out a display window or a group of window lights to vent a store fire. Before doing so, truck crews must be sure they are not dealing with a smoldering fire, which has a potential for backdraft. If no fire is showing and the store is heavily charged with smoke, or if the display window is hot to the touch, the roof should be opened before the display window is knocked out.

Short ladders, if handy, are useful in knocking down ceilings after the roof has been opened for venting, for punching out the cockloft enclosure under a skylight, or for knocking out a transom or glass panel below a skylight. Ladders also are useful for knocking down ceilings that are too strong to be knocked down with pike poles, especially when the operation would require more than one fire fighter. Ladders also can be used to knock down the ceiling over a basement that is being vented through the walls beneath display windows.

ADVANCING HOSELINES

A number of situations require the use of ground ladders to advance hoselines into a building. For example, occupants might be leaving by stairways at the same time engine companies must get lines in to attack or head off the fire or to protect escaping victims. If the stairways are not well located, engine crews must use ladders to get into good positions to attack the fire (Figure 9.9). (Sometimes there are no stairways at all at the rear of a building.) Also, stairways, although free of people, might be damaged, or there might not be enough of them.

When interior stairways and corridors are available, they will normally be used to advance initial attack lines to ensure control of these passageways. If additional lines are advanced through the same stairways, they could become overcrowded and thus unsafe for fire fighters. Ground ladders provide an obvious means of relieving such conditions.

Placing Ladders

Ground ladders should be placed where they will be of use in the overall fireground operation — for advancing initial attack lines, backup lines, or lines that might be used to cover interior and exterior exposures. Initial size-up by truck officers will indicate where engine companies may have to enter the building. Ladders should be raised at these positions on arrival. That

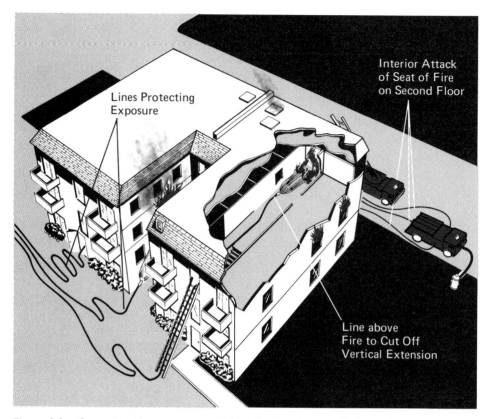

Figure 9.9. Ground ladders can be used for effective placement of hoselines.

is, truck crews should raise ground ladders at strategic points without waiting for orders from the chief officer.

Ground ladders should not be raised simply as an exercise, since this could delay other important truck operations. It is wasteful to raise a ladder where it obviously will not be needed, but it is just as wasteful not to raise a ladder where it might be needed. Occasionally, a properly positioned ladder will not be needed by engine crews, possibly because the fire was quickly controlled from other positions, but this should not deter truck companies from placing ground ladders in similar positions at future fires where the ladders could be crucial.

After ladders have been raised for rescue, advancing lines and other operations, a detached building can be ringed with ground ladders. An attached building can be laddered only at the front and rear. If at all possible, truck crews can avoid placing ladders in front of building entrances, where fire fighters entering the building, or occupants leaving it, could inadvertently move it out of a safe position or knock it down. The ladder might also interfere with traffic moving through the entrance. If handlines become tangled in the butt, attempts to free them will delay fire attack and could require the ladder to be repositioned.

Raising Ladders from Roofs

Ground ladders can be raised to the sides of a tall building from the roof of an attached or nearby lower building. Truck crews should first raise a ladder to the roof of the lower building, climb up to the roof, and then get additional ladders up. If the lower building is not more than three stories high, additional ladders can be lifted up by crews on the ground to truck members on the roof, who then should be able to get hold of the ladders and pull them up. If the lower building is more than three stories high, the ladders should be hoisted up by rope.

Once additional ladders are on the roof of the lower building, they are raised to the proper positions on the side of the taller building. Sometimes this will require ladders longer than those used to get to the roof. This would be the case, for example, if ladders were being placed to the sixth floor of the fire building from the roof of an attached two-story building. Truck crews should know the height of the required raise and which ladders will reach. It is not always the shorter ladders that are raised from a roof.

Climbing Ladders

Normally, fire fighters carrying lines should be at least 20 feet apart as they climb a ground ladder. If the ladder was raised at a less than normal angle, this distance should be increased. This does not mean that the crew should hold the hoseline at 20-foot intervals. Rather, the hose should be grasped at the couplings — that is, at 50-foot intervals. As the fire fighters climb, they should allow loops of hose to hang off the side of the ladder. This will distribute the total weight of the hose properly along the ladder.

Once the line has been advanced into the building, it should be moved clear of the ladder so it hangs down the face of the building. The ladder will then be freed for other uses, and hose movements will not endanger fire fighters on the ladder.

POSITIONING FIRE FIGHTING STREAMS

Ground ladders can be used to hold and position streams being directed into a building. There are two methods of placing ladders for this purpose.

In the first, the ladder is positioned and used to knock out a window. The topmost three or four rungs are allowed to extend over the sill and into the window as the window is being removed. The line is then carried up and tied to the ladder so the stream will be directed through the window opening. The nozzle is opened and the stream directed onto the fire (Figure 9.10).

In the second method, the ladder is raised over the window with the tip placed against the wall above it. Sometimes the tip is placed on the sill or into the next higher window. Again, a line is carried up, tied to the ladder, and directed onto the fire.

In either case, the ladder should be tied to the building if at all possible. If the ladder is in a window, a tie-in with a rope hose tool will suffice. If the ladder is against a wall, a pike pole can be placed across the inside of a lower window, tied to a rung, and pulled tight against the inside wall. Whenever possible, the ladder should be braced by fire fighters at the butt, whether tied in or not.

The hoseline should be tied to the ladder with a rope hose tool or strap in a manner that will allow the rope to absorb some of the reaction of the line when the nozzle is opened. It should, therefore, not be tied rigidly to ladder rungs; rather, it should be suspended between rungs by the rope or strap. One way to do this is to hold the nozzle between two rungs, pass the rope behind the rung just above the nozzle, wrap it around the rung to use up the slack, and then hook it to the second rung up. With the hose suspended from the rope, it can move back when the nozzle is opened, but it will not exert a strong pull on the ladder. The ladder will remain in place.

Streams directed from ground ladders can be effective in knocking down fire at various points in a building. They also can be used to knock down the

Figure 9.10.

fire ahead of engine crews advancing with interior lines, especially in rooms flanking the corridor. The ladder streams will help with interior operations if the building is properly vented. As interior lines approach ladder streams, the latter should be shut down; both types should not be used together in a large open area, unless the interior lines are not advancing toward the ladder streams.

Once a ladder stream has knocked down the fire in an area, it can be shut down, untied, advanced, recharged, and used for interior fire attack. However, fire fighters must use care once the stream is shut down. Any steam or smoke in the area should be allowed to clear; then the condition of the interior, especially the floor, should be checked. Before they enter the building, fire fighters should make sure that the fire did not flare up again after the line was shut down.

If it is safe to enter the building through the window, the line should be untied from the ladder and placed over the sill. The nozzle operator should enter the window only when a second fire fighter has climbed to a position immediately outside the window, to assist in advancing the line and back up the nozzle operator if fire erupts in the area.

OTHER USES

Several other uses for ground ladders were mentioned at the beginning of this chapter. The movements involved in these uses are, for the most part, obvious.

Forcing Entry

A ground ladder can be used to force an inward-opening door. The butt of the ladder is placed against the frame of the door, near the lock, with the rungs vertical. With two or three crew members on each side, the ladder is pushed against the door. The rungs are used as handholds, and the door is pushed in — not battered — to break the lock away from the jamb.

Display-window enclosures, partitions, paneling and other such building features can be pushed away in a similar manner.

Ground ladders also are used to knock out large windows. The length of the ladder allows fire fighters to maintain a safe distance from the shattering glass.

Transporting the Injured

Short ground ladders are useful as litters for lowering or carrying injured victims from buildings. For lowering, the ladder should be rigged as a litter (see Chapter 6) and lowered with an aerial unit or with rope.

Covering Weakened Areas

Ground ladders sometimes are placed over building features weakened by fire or by fire fighting operations, to allow safe passage. For example, a stairway damaged by fire might still be utilized if a ground ladder is laid over the stairs, with the butt and tip in contact with firm flooring or undamaged stairway supports.

A ground ladder also can be placed over a weakened or suspect area of a roof or floor to allow crews to work safely in that area. Each end of the ladder must be supported by a solid part of the roof or floor.

SUMMARY

Ground ladders serve many purposes in fire fighting operations. Regardless of how the ladder is used, it must be handled safely. It should be carried, spotted, raised, pivoted, extended and placed with regard to its nested and extended length and its overall bulk and weight. In addition, overhead obstructions must be observed and avoided. Truck companies should practice the under-wire raise with all sizes of ladders.

Training with ground ladders is essential if truck company personnel are to perform efficiently on the fireground. In training sessions, ladders should be handled by the same number of fire fighters as will be available on the fireground — not by larger-than-normal crews. When help will be required to raise the larger extension ladders — for example, 40- to 50-foot pole ladders — engine personnel should be trained to render such help. However, once the ladder is in position on the fireground, the extra personnel should resume their assigned duties.

SALVAGE   10

Water is the prime extinguishing agent used by fire companies, and it will probably remain so for some time to come. Fire departments are organized and trained to apply water onto and into structures. Streams are needed to protect victims until they can be rescued, to protect exposures, and to contain and extinguish fires. However, the same water used to limit the damage done by fire can itself damage a building structurally and ruin its contents. Salvage operations can reduce water damage in almost any structure or occupancy.

There are two types of salvage operations — those that protect the contents of a building, and those that protect the building itself from structural damage due to the weight of the water. The first mainly entails the proper placement of salvage covers to protect contents and catch the water (Figure 10.1A). The second requires that water be removed from the building before it overloads the structure (Figure 10.1B). The two salvage operations are equally important in reducing property losses.

In recent years, salvage has been referred to as the "lost art" of the fire service. In many cases, salvage operations are neglected because of a supposed lack of personnel. A review of such cases often indicates that truck crews could easily have carried properly folded salvage covers into the building and placed them while on their way to performing other duties. Fire fighters awaiting assignments at the scene also could have been engaged in salvage operations. Unfortunately, once a fire department or fire company believes it does not have the personnel for effective salvage work, that work will not be done, regardless of the actual personnel situation.

It makes little sense for fire fighters to put all their effort into controlling a fire that may do a few hundred dollars' worth of damage while allowing water to ruin thousands of dollars' worth of office equipment or domestic furnishings. Yet that is very often the case.

Another detriment to proper salvage is the misconception that salvage is related to overhaul. This idea is perpetuated by such items as "Salvage and Overhaul" manuals, whose very titles indicate a connection between the two operations. Actually, they are no more related than any other two truck

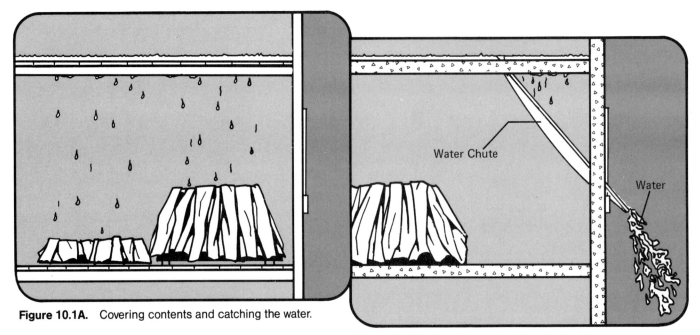

Figure 10.1A. Covering contents and catching the water.

Figure 10.1B. Removing water from the building.

operations. They have different objectives, are initiated at different times, and require different procedures:

- The main objective of salvage is to protect the building and its contents from water damage; the main objective of overhaul is to make sure the fire is completely out.
- Salvage operations should be started as soon as fire attack begins, or as soon thereafter as possible; overhaul operations are not started until the fire is apparently extinguished.
- Salvage operations are performed with salvage tools, including salvage covers, conduits, chutes, submersible and portable pumps, venturi siphons, drain screens, squeegees, mops, brooms, smoke deodorants, and allied equipment; overhaul requires truck tools and hoselines.

This chapter deals with salvage operations, and overhaul is discussed in Chapter 12. It is important that truck crews be thoroughly trained in both operations, so they realize the differences and do not neglect either operation on the fireground.

The type of salvage operations required in a particular building will depend to some extent on its construction. For example, the floors in a fire resistant building are made of concrete, which usually will hold the water directed onto it. Salvage operations in such a building primarily require containing the water, directing it along preset paths, and removing it through drains, windows, stairways or chutes. However, some water might seep down to lower floors through cracks in the cement floors, where pipes pass through the floor, and through stairways. Therefore, the contents of lower floors also must be protected (Figure 10.2).

Water will seep quickly through wooden floors (Figure 10.3). Because of this, in buildings with wooden floors the first salvage operations must be directed toward protecting the contents of the floor below the fire floor,

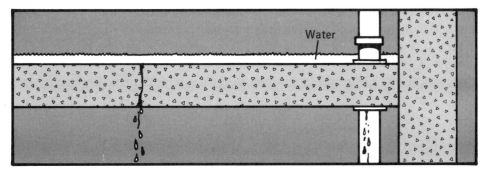

Figure 10.2. Concrete floor holding the water. Some seepage occurs through cracks and pipes passing through the floor.

mainly with salvage covers. However, it is also necessary to remove the water from such buildings to keep it from seeping down from floor to floor.

Size-up of the fire situation will indicate which salvage operations should be initiated first, and where. To be effective, salvage should begin along with fire attack. That is, the building and its contents should be protected from water damage when water is first directed into the building. There should be no hesitation in calling for extra companies to perform salvage operations when it is obvious that uncontrolled water damage could be extensive.

PROTECTING BUILDING CONTENTS

Building contents are protected mainly by covering them to keep them from being damaged by water and debris. The flow of excess water should be directed away from stock, furnishings and equipment. Also, "catchalls" — tublike affairs — can be set up to catch water dripping from above. Salvage covers are an important item in all three operations.

Salvage Covers

Salvage covers are essentially large sheets of waterproof material. They are available in several sizes, materials and shapes. Some are fire resistant, and some are not. Some salvage covers must be placed with a particular side open

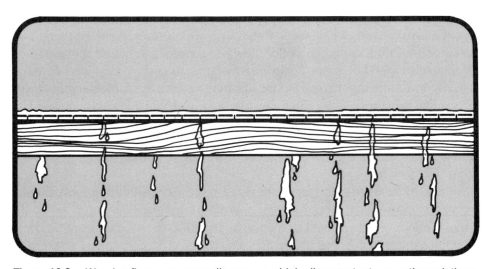

Figure 10.3. Wooden floors are generally porous, which allows water to seep through them quickly.

to the water. Fire fighters should be familiar with the type of salvage covers carried by their company, since the type often determines how the covers can be used. It is equally important that salvage covers be purchased with regard for the average number of personnel responding with the company. The size and weight of a cover affect the number of people required to place it.

Various ways of folding salvage covers for transporting them on the apparatus are given in manipulative-skill manuals. The appearance of the folded cover should indicate whether it can be spread by one person, or whether two or more people are required. If it does not, then the cover should be tagged to indicate the number of personnel required for its use, and any special use for which it has been designed or folded.

The types of salvage covers used, the folding methods, and the carrying and spreading techniques should be standardized within a department and, if possible, with neighboring departments. This will allow covers to be used interchangeably and spread quickly so salvage crews can be released for other duties as required by the fire situation.

Covering Building Contents

Salvage covers should first be spread over the building contents that are in the most danger of being damaged by water. In most cases, these are on the floor below the fire floor, directly under the fire — that is, under the area to which the streams will be directed. These contents will be subjected to water seeping from above. Once the contents of this area are well protected, covers should be placed on the contents of surrounding areas.

In rare cases, the area under the fire contains few items that would be damaged by water. In this instance, either the contents of surrounding areas should be covered first, or — if the fire floor contains equipment or furnishings that are sensitive to water damage — its contents should be covered first. The important point is to first cover the items that could suffer the most water damage because of either their position or their value.

Besides being spread to protect items resting on the floor, salvage covers can be rigged over shelves mounted on walls (Figure 10.4). They can be nailed to the wall above the shelves through grommets provided in the covers and draped to cover the shelves. For this use, hammer-and-nail kits should be part of the salvage equipment carried by truck companies.

Salvage covers also should be spread over the contents of the fire floor and floors above the fire, when necessitated by fire fighting operations there. For instance, before a ceiling is pulled down to search for the lateral spread of fire, covers should be placed over nearby items to keep debris (and water, if used) from damaging them. The fire situation and the available personnel will determine whether or not such salvage operations are feasible, but crews entering an area with pike poles to pull ceilings can easily carry and spread covers if the fire situation is not too severe. When covers are not available, unnecessary damage can be avoided by moving the room's contents away from the work area.

When the number of salvage covers is limited, the available covers should be used to protect the most valuable contents. Every salvage cover should be used as effectively as possible. One way to do this is to move the contents into as small a space as possible before covering them. If the contents of an average room of a dwelling are piled together, they can usually be fully protected with two or three salvage covers.

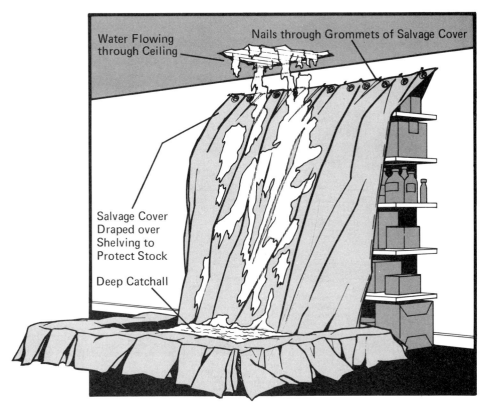

Figure 10.4. Salvage covers can be used to protect stock on vertical shelving.

Controlling the Water Flow

Salvage covers can be used to control the flow of water along a floor. Placed in doorways or stuffed under doors, they will block the movement of water out of the fire area or the area below the fire. (In the latter case, the contents should previously have been protected with salvage covers.) However, water that is blocked and left to accumulate will soon begin to leak down through the building. It must therefore be removed down stairways or other openings. (These operations are discussed in more detail in the next section.)

Salvage covers can be folded for use as conduits to direct accumulated water to stairways and then down the stairways and out through exterior doorways. Salvage covers to be used as conduits should be prefolded at the station, rolled up, tied, and carried on the apparatus ready for use. This saves time that would otherwise be needed to rig them on the fireground.

To make a conduit, two salvage covers are laid out flat, end to end, with some overlap. The sides of both covers are then rolled together toward the center, to form one narrow length. This is then coiled up tightly from one end, and tied securely.

To use the conduit, fire fighters carry it to the top of the stairs, unfasten it, unroll it down the stairs, and spread the sides apart. The rolls along the edges keep the water within the conduit. Water can be guided to the conduit with squeegees and brooms, or through a second conduit (if required) placed between the accumulated water and the stairway.

Conduits are also used in two ways to guide water along floors. On watertight floors, they can be unrolled to form a guide or wall to keep the water confined as it is moved toward stairs or a drainage system (Figure 10.5). A pair of conduits used in this manner can be very effective. On porous floors or

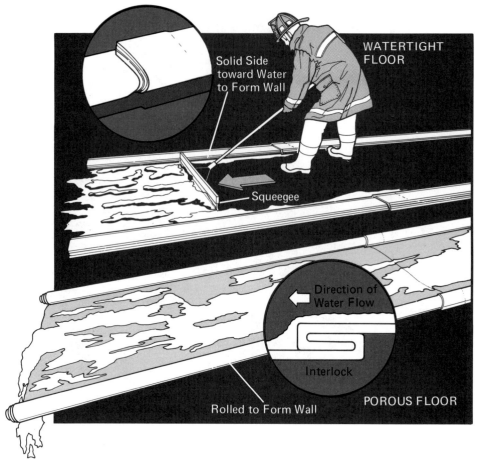

Figure 10.5. Conduits can be used as guides for water over watertight floors and to carry the water over porous floors.

those that are not watertight, the conduit is unrolled and spread flat, except for a roll on each side to hold water. The water is moved over the conduit in the desired direction.

Care must be taken, when conduits are made up and placed, to make sure that the interlock works effectively. The cover forming the upper part of the interlock must be placed toward the flow of water ("upstream"), so the water will pass over it and down onto the next cover.

Sawdust, carried in canvas bags with tubelike openings, can be spread in small or confined areas to absorb water. Sawdust also can be used to guide and control the flow of water away from areas in which it could do damage.

For the latter use, two parallel beads of sawdust are laid along the path the water should take. The water is pushed through, between the beads, with brooms or squeegees. Move water along gently at first, to avoid breaking up the sawdust guides. Once the sawdust becomes soaked with water, it will tend to stay in place unless the water is moved too quickly.

Such sawdust guides can be used alone, to direct water to a salvage cover conduit, or to continue the path at the outlet end of a salvage cover conduit. In all cases, the water should enter between the beads so it will not destroy them.

Catchalls

Salvage covers can be rigged as basins, generally referred to as "catchalls," to catch and hold water dripping from a ceiling. The cover can be rolled in

Figure 10.6. A shallow catchall can be made quickly by rolling in all edges of a salvage cover.

quickly from all edges to form a flat, shallow catchall (Figure 10.6); or draped over four ladders or other suitable material to form a deep catchall (Figure 10.7). Because they are small, catchalls cannot be used as the sole water-damage control device when extensive fire fighting operations are in progress. They are most effective at comparatively small fires or in outlying building sections at larger fires.

Catchalls are, however, effective in keeping moderate amounts of water off building contents, and in preventing water from moving around the floor or seeping down to lower floors. Often, fire fighters use pike poles to punch holes in the ceiling where the water is dripping (Figure 10.8). This helps drain water and prevent the collapse of a big piece of the ceiling which would splash water over a large area. Once the catchall is filled during these operations, it must be dumped carefully so the water does not spill over the floor and pass through to lower stories.

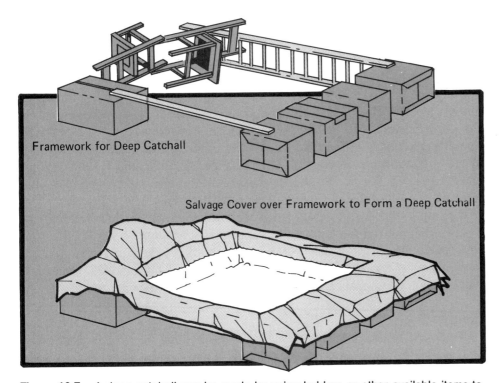

Figure 10.7. A deep catchall can be made by using ladders or other available items to support a salvage cover.

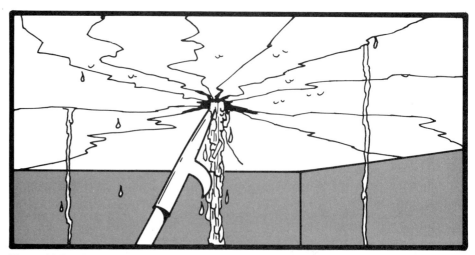

Figure 10.8. Converging cracks in ceiling indicate accumulation of water. To relieve water, gently poke a small hole where cracks converge.

If the full catchall is too heavy to move, portable pumps can be used to remove water. Care should be taken to make sure the connections on the hoselines carrying water out of the building are tight and do not leak.

REMOVING WATER FROM BUILDINGS

A lot of water can collect within a structure while an extensive fire is being fought. Although salvage covers might be placed below and around the fire, collected water will run down through the structure. This damages uncovered contents of lower floors, becomes deep enough in spots to damage items resting on the floor, and, most important, will add weight to the floors of a structure already weakened by fire. Truck company personnel must remove water before it can do such damage or cause the building to collapse.

The sooner the water is removed, the less damage it will do to floors and carpets and the less chance there is of leakage to lower floors. Often, there will be little or no permanent damage to flooring materials or to ceilings below floors if the water is removed quickly and the floors allowed to dry out.

One particularly effective way to remove water from smaller buildings is to quickly push it out with squeegees or brooms. Where the water is to be pushed through a hallway, stuff salvage covers under doors and place them along wide openings to rooms to keep the water from entering dry areas (Figure 10.9). When this operation is started early enough, there should be little or no leakage to lower floors. The effectiveness of the operation also depends on the flooring material and the quality of its construction.

A number of other ways to remove accumulated water from a building are discussed in the remainder of this chapter. Most are more complex than the squeegee operation, but they may be necessitated by the construction of the building or by the amount and location of water to be removed. Usually, a combination of several methods will be required to "dry out" a building.

Chutes

The use of rolled salvage cover conduits and sawdust has already been discussed. In addition, more permanent chutes can be made from salvage cover material. These chutes can be used to direct water from one floor out through the windows (or other openings) of the floor below.

Each chute is made from a strip of salvage cover material about 10 to 12 feet long and a pair of wooden poles or aluminum pipes of the same length. The long edges of the strip are rolled and fastened around the poles. The chute is rolled, pole to pole, into a compact unit for transportation. For use, it is unrolled, rigged below a hole in the floor, and placed so it angles down to a window on the next lower story. The chute allows water to escape directly from the floor above with minimum movement through the building.

The chute can be rigged below the hole on a short straight ladder or an A-frame type ladder. The ladder should be tall enough to hold the upper end of the chute close enough to the hole to keep the water from splashing down. The chute poles can be tied to the ladder with short pieces of rope provided on the chute. Salvage covers should be spread over items close to the chute to prevent damage if there is a spill.

The lower end of the chute should extend far enough out the window so no water will fall back into the building. Windows below should be closed, or covered if broken. Personnel should be warned to stay clear of the area.

Figure 10.9. Salvage covers help block water from flowing into dry areas.

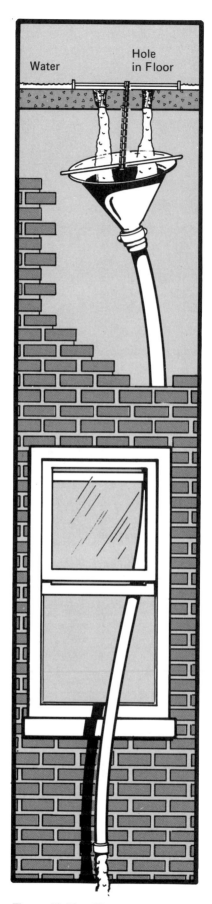

Figure 10.10. Chutes can be used to quickly and efficiently remove water from a building.

Another effective chute can be fabricated from a funnel-type device and an old hose. One chute of this type is illustrated in Figure 10.10. To use this chute, truck crews cut a hole in the floor (the water is blocked with a salvage cover while the hole is being cut). A bar is laid across the hole to support the chute, and the chute is positioned. The hose is run out a window or into the drainage system. The salvage cover is then arranged to form a U-shaped dike to direct the water toward and into the hole. The water can be pushed to the hole with squeegees. A drain screen can be constructed as part of the top of the chute or placed in the chute when it is positioned in the hole.

Drains

Floor and wall drains built into the building can be used if they are located fairly close to the accumulations of water. (Water should not be moved very far across floors if this can be avoided.) In any case, large quantities of water should not be moved to small drains since, if the capacity of a drain is exceeded, another accumulation of water will be formed. Also, the drain must be kept free of debris so that water does not back up in the area. In most situations, built-in drains must be used in combination with other water removal methods. Locations of adequate drains should be determined during prefire planning surveys.

Floor drains might be located throughout a building or only in certain parts. They usually have some sort of a screen-type cover. Some drain directly into sewers and others into dry wells. They are ideal outlets for water because they are designed for the purpose. Little effort is involved in their use, except to direct water to them and keep the covers free of debris.

In some buildings, wall drains (scuppers) have been placed at the base of the walls even with the floor. The interior opening is usually covered with a heavy wire mesh; the exterior opening usually has a hinged cover. Scuppers are also ideal for removing water and are used in the same way as floor drains.

Toilets

When a toilet is unbolted from the floor and lifted out of place, a sewer-pipe opening is exposed. In most cases, the opening is 4 inches or more in diameter and can be used to remove much water quickly from nearby areas if the drain is kept from clogging. The flooring in bathrooms is usually the most water resistant in a building, so this operation can be very effective. It should be repeated as necessary, from story to story, when water has accumulated throughout a building.

The sewer pipe is not usually sealed to the floor around it, so water can drop down to lower stories through the space between the pipe and the floor. To prevent this, the space should be closed up as much as possible. Anything that can be stuffed around the pipe will hinder the flow of water into the opening. A salvage cover can be placed on the far side of the pipe to keep water from flowing past it, and part of the cover stuffed in around the pipe. Ceilings below the toilet should be checked for leakage.

To prevent the sewer pipe from becoming clogged with debris, a drain screen should be placed over the opening. An effective screen can be made up of 1-inch mesh on a 10- to 12-inch steel ring or square frame.

Sewer Pipes

In some structures, mainly commercial and industrial buildings, sewer pipes are exposed rather than covered by partitions. They usually run along

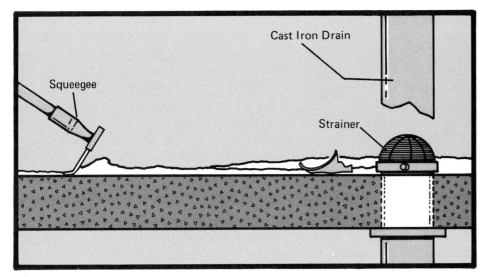

Figure 10.11. Exposed sewer pipes can be effectively used to remove water.

one wall of the building. In stores, they are most often located in the rear work areas along with other utilities. These sewer pipes provide another effective means for removing large quantities of water. However, because the sewer pipes must be broken to be used for water removal, small amounts of water should be removed by other means, and this method should be used only to remove large quantities of water from an area near the pipes.

The sewer pipe is knocked open at floor level or as close to the floor as possible (Figure 10.11). Any pieces of pipe that extend above the floor are broken off to make the opening even with the floor. Sewer pipes in older structures are usually made of cast iron, while in newer structures they can be cast iron, plastic, or a ceramic material. Any of these can be easily opened with forcible-entry tools. A drain screen should be placed over the opening to keep out debris.

This operation can be repeated on several floors of the building, as necessary. However, the sewer pipe on the uppermost floor must be opened first, and that floor drained of water, before the pipe is opened on a lower floor. If the sewer pipes on two floors are opened at once, water from the upper floor will pour out onto the lower floor.

A building can usually be drained through the sewer pipe on the fire floor, especially if it is opened soon after fire attack begins. However, there should be no hesitation in opening the sewer pipes on lower floors if such action is required.

Openings in Walls

When large quantities of water are being used to extinguish a fire and water is accumulating quickly, large openings may be required to get the water out and keep the building safe. One such opening could be made by removing the wall immediately below a window, from the sill to the floor. Hand tools, power tools, or both can be used, but truck crews must be careful not to cut or damage structural members while removing the wall.

If possible, the windows below the opening should be closed or covered before the opening is made. Personnel below the opening should be warned to stay clear of falling debris and water.

This is an extreme action that should be used only under extreme conditions — when large amounts of water endanger the building and the fire fighters operating within it. In most instances, the wall cut will first be used to remove water from the floor below the fire while streams are still being directed onto the fire. Truck crews engaged in such an operation must be kept informed of the fire situation above them and of the amount of water on the fire floor. They must also be on the lookout for signs of building collapse. Portable radios should be used for quick communications concerning the water removal operation and the condition of the building.

Once engine companies have extinguished the fire near the windows on the fire floor, similar openings might have to be made there. With these two floors opened, the greatest accumulation of water can be removed.

Pumps

Various types and sizes of portable pumps, powered by electricity or gasoline, are available for removing water from buildings. They can be used alone or in combination with other water removal methods. However, most of these pumps have small capacities so cannot be used for the quick removal of an appreciable amount of water.

Pumps can be effective in removing water from areas of a building where other means are not feasible. Among these areas are basements, elevator or other shaft pits, and other low areas without drains.

Elevator pits and large utility shaft pits can be used to quickly remove water from upper floors. However, this operation is effective only when the bottom of the pit is below the level of the basement floor. Water from above is directed down the shaft, and submersible pumps, portable pumps, or siphons are used to remove the water from the pit to the outside. Water can be removed from basements in the same way.

Because of the great amount of debris in the water, first line pumpers should not be used in dewatering basements or pits. Venturi siphons will get the water out faster than drafting with a pump, and without the danger of damaging fire apparatus.

Here are a few words of caution about flooded basements. Fire fighters should remember that although water always shows a flat surface, the floor beneath may not be flat. For instance, there could be more than one level of basement flooring (Figure 10.12). Fire fighters must be very careful when entering a flooded area and should probe ahead with pike poles to determine

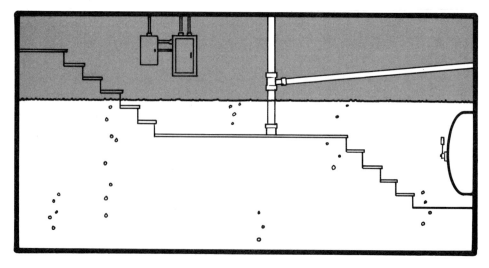

Figure 10.12. When water fills basements, fire fighters should be extremely cautious. Water shows as a flat surface; however, the floor surface beneath the water may not be level.

Figure 10.13. To avoid being pulled beneath the surface by the suction, do not clear clogged pipes or open drains by reaching into them. Instead, clear the clog with the end of a tool.

the location of the floor, stairs and other obstacles. Many fire fighters have been injured, and a few drowned, through lack of caution when working in such areas.

Also, the water in a basement can get deep. Fire fighters should not put their faces under water, or even close to the water surface, to open a clogged drain (Figure 10.13). The suction that results when the drain is opened can pull under even the strongest fire fighter. Clogged drains must be cleared with the end of a tool.

SUMMARY

If salvage operations are to be effective, truck company personnel must know the reasons for these operations and how to work most effectively in accordance with the fire situation. In particular, they must know how and where to place salvage covers to protect building contents, and how to use them to control water flow and move water out of the building.

Truck personnel must also be familiar with the use of chutes and building features for water removal. The choice of methods should be based on the amount of water in the building and the fire situation.

The safety of fire fighters engaged in salvage and other operations must be taken into account. Those involved in salvage work must be cognizant of the fire above and of how it might affect their work. They must also be aware of how their work could endanger other fire fighters in and around the building.

ELEVATED STREAMS   11

One of the most important uses of an aerial unit is to provide an elevated stream for fire attack or exposure protection. The elevated stream is directed to the fire or exposure from atop the raised aerial unit. In most cases, the stream will be developed by a heavy stream device — that is, a ladder pipe or platform pipe — and most of this chapter is devoted to such operations. Occasionally, a handline will suffice for elevated stream development; the last section of this chapter discusses the operation of handlines from aerial units.

Elevated streams can be effectively directed into or onto the upper parts of tall buildings, beyond the reach of ground-mounted or engine-mounted heavy stream devices. They are also useful on large, sprawling structures, outside storage yards (such as lumber yards and oil storage tank farms), piers, ships, and, in general, areas where their height and reach provide access to the fire or exposure. To maximize this effectiveness, a ladder or platform pipe should be equipped with solid-stream nozzles of several diameters and at least one large-caliber fog nozzle. This will allow fire fighters to develop the proper stream for any combination of fire situation, weather conditions and water supply.

SETTING UP THE AERIAL PIPE

Ladder and platform pipes can, as noted, be used for direct fire attack, exposure protection, or a combination of both. If possible, the apparatus should be positioned with regard to both the wind direction and the location of the fire and exposures. A water supply must be developed and connected to the aerial pipe. Finally, the pipe must be rigged, raised and charged.

Spotting the Turntable

In some situations, an aerial pipe will have to be operated with the truck in an unfavorable position. For example, the turntable might have been spotted for rescue or for venting while handlines were being used for fire attack. Continued growth in the size and intensity of the fire could have forced fire fighters to abandon the handlines and place the aerial pipe in service. Owing to the placement of pumpers and other apparatus, it may be impossible or impractical to reposition the truck; the pipe would then be operated as effectively as possible from its original position. If it is feasible to quickly move the truck to a more advantageous position, this should be done.

However, in some situations aerial pipe operations will begin upon arrival on the fireground. Then the turntable should be spotted for maximum effectiveness.

Buildings. Where wind is not a factor, the turntable should be spotted for maximum coverage of the fire area. Usually, this will be at the center of the fire building (Figure 11.1). If the building is fairly wide, the unit should be spotted in the middle of the involved area.

When the wind is blowing across the face of the fire building so that nearby structures are exposed, the turntable should be spotted between the fire and the exposures (Figure 11.2). This will allow the elevated stream to be directed to the fire building and, at the same time, to be in position to cover the exposures in case there is danger of further fire extension.

Open storage areas. Here, again, when wind is not a factor, the turntable should be spotted at the center of the fire area for maximum coverage of the fire and nearby exposures.

However, when there is a wind — especially a strong one — aerial turntables should be spotted at the flanks of the fire between the main body of fire and exposures (Figure 11.3). The units should not be positioned directly in the path of the fire. The elevated streams will thus be in position to be directed onto the fire and the exposures, and water spray caught by the wind will fall on the exposures. At the same time, the aerial units will not be endangered if the

Figure 11.1. If wind is not a factor, the turntable should be at the center of the building or fire area.

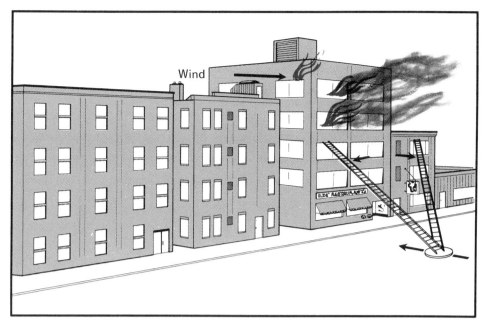

Figure 11.2. When wind is blowing across the face of the building, the aerial unit should be placed downwind in order to protect exposures and operate on the fire building.

elevated streams cannot control the fire. (They would be endangered if they were spotted directly in the line of fire travel in an attempt to beat back the fire but failed to do so.)

Elevated streams are also used to protect fire fighters advancing handlines toward a fire in an open storage area. This operation can be especially important when the radiant heat from the main body of fire is intense, or when the lines must be advanced around piles of stored material. The turntables

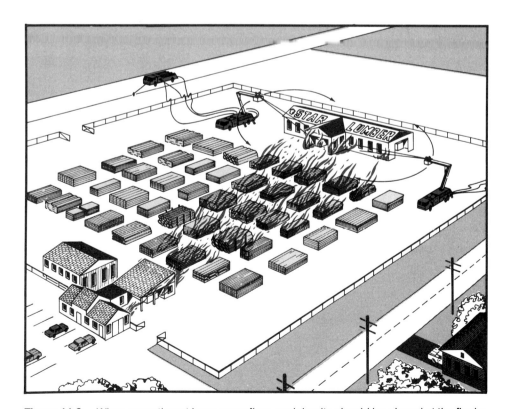

Figure 11.3. When operating at large open fires, aerial units should be placed at the flanks of the fire, if possible, to use streams to protect exposures and hit the fire.

should be spotted either behind or at the flanks of the advancing fire fighters. The aerial ladder or platform should be raised over the crew and streams directed just ahead of them as they advance.

Flammable liquid handling facilities. The storage area positioning discussed above should be used at facilities where flammable liquids are processed or stored. However, the turntable should never be spotted in line with either end of a horizontal tank. This is an extremely dangerous position, since the ends of the tank will blow out if the tank explodes. If possible, the unit should be positioned at the side of the tank, and the elevated stream used to cool the tank as well as to attack the main body of fire (Figure 11.4).

Developing the Water Supply

The aerial-pipe water supply is most often developed in cooperation with engine companies. Engine crews usually obtain the water (from hydrants, fireboats or drafting sources), lay the supply hose up to the aerial unit, and provide the pumping pressure. Truck crews then are responsible for moving the water from the aerial-pipe siamese connection to the pipe.

For safety and efficient movement of the water, the pumper should be positioned close to the truck — between 50 and 100 feet from it. Where fire-to-hydrant layouts are the usual department procedure and the aerial unit does not have its own pump, a pumper at the hydrant should deliver water to a pumper close to the aerial unit. This operation can be modified where short lays and large diameter hose — above 3 inches — are used. Two-pumper

Figure 11.4. Elevated streams can be effective at fires involving flammable liquids. Where tanks are involved, care must be taken to avoid placing personnel or apparatus at the ends of horizontal tanks.

operation is also most efficient with hydrant-to-fire layouts, but it may not be necessary for short lays of large diameter hose or dual layouts of 2½- or 3-inch hose.

The water supply operation will be effective only if engine and truck crews are able to work well together. Joint operations should be practiced on the training ground, where fireground problems can be duplicated and where time is available to evaluate and solve these problems. Mistakes made on the training ground are not costly; they *are* at the fire scene. Moreover, as training progresses and skills improve, engine companies will become aware of the problems of truck companies, and vice versa. When each group is able to anticipate the movements of the other and operate accordingly, the net effect will be increased efficiency in all joint fireground operations.

Rigging the Aerial Pipe

Platform pipes and ladder pipes should be rigged according to manufacturers' recommendations. In addition, truck crews must follow manufacturers' suggested limits as to angle, extension, elevation and water flow rate in gallons per minute.

Platform pipes. A platform pipe is permanently mounted to the aerial platform, attached to the supporting mechanism. To place the pipe in service, the water supply lines are connected to the intake siamese, the platform is raised, and the pipe is released, aimed and charged.

For safety, the pipe should be charged before the basket is moved toward the fire. The platform pipe stream and the spray system under the basket will then be available to protect fire fighters in the basket (and the apparatus) if fire should issue toward the unit (Figure 11.5). In addition, each fire fighter in the basket should wear a life belt connected to a side rail.

Bed-mounted ladder pipes. A ladder pipe can be permanently mounted on the bed section of the ladder or placed on the tip of the ladder on the fireground. To place a bed-mounted pipe in service, the supply lines are connected to the intake siamese, and halyards are attached (if not kept attached) to the pipe for its vertical control by a fire fighter on the turntable. (The turntable operator provides lateral movement by rotating the turntable.) Finally, the nozzle is released from its holder, the ladder is raised to position, and the pipe is charged.

The turntable should be rotated in accordance with the manufacturer's recommendations. Some manufacturers suggest that the hand crank be used when the pipe is in operation; others recommend hydraulic rotation. However, the hand crank should be used whenever the turntable cannot be rotated smoothly with the hydraulic system.

Ladder-tip pipes. Several additional steps are required to rig a ladder pipe to the tip of the aerial ladder once the apparatus has been positioned. Consequently, it takes more time to get this type of ladder pipe into service.

The first step is to place the pipe on the ladder, with the proper nozzle for fire and wind conditions. It must be positioned dead center and locked down securely with whatever locking arrangement is provided by the manufacturer. For additional safety, the pipe can be tied to the ladder rungs with a ladder rope, rope hose tool, or leather or fiber strap. Next, the ladder pipe hose is connected to the pipe (if it is not preconnected).

Figure 11.5. Before the platform is moved in toward the fire, the platform pipes should be charged.

At this time a decision must be made: will the pipe will be controlled by a crew member on the ladder (directly with the handle) or by one on the turntable (with halyards)? The handle, with or without halyards, should be attached now. Then, a siamese is connected to the ladder pipe hose (if not preconnected), the supply lines are connected to the siamese, the ladder is raised, and the pipe is aimed and charged.

The siamese should be equipped with a pressure gauge. There should be a shut-off valve between the siamese and the ladder pipe hose if the siamese is not gated. These two devices will allow control of the water supply so that excessive pressure is not applied to the pipe.

If the pipe is to be controlled at the tip of the ladder, the fire fighter should be locked in to a rung (not side rails) with a life belt (Figure 11.6). In addition, footplates should be provided near the top of the fly section for the pipe operator. However, a leg lock should not be used on the ladder; the combination of life belt and footplates will allow the fire fighter to work safely while using both hands to control the pipe.

The ladder should be elevated to a comfortable climbing angle and extended to clear the footplates on the fly section. The ladder pipe operator then takes position on the ladder and is raised safely with the ladder. This is less time consuming than raising the ladder and then requiring the operator to climb into position. (It is important that this movement be practiced before it is attempted on the fireground.) As with platform pipes, the ladder pipe should be charged before it is moved in toward the fire so the operator will be protected by the stream.

The pipe operator on the ladder provides only the vertical control of the nozzle; the ladder must be rotated to move the nozzle laterally. However, some ladder pipes are designed to be moved horizontally 15 degrees to the right or left, while larger lateral movements require rotation of the ladder. The manufacturer's recommendations must be followed in all cases.

The pipe operator and the aerial ladder operator on the turntable should communicate through the aerial intercom system for smooth operation. If the

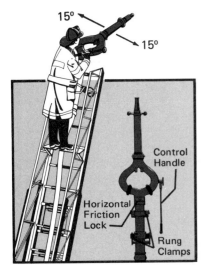

Figure 11.6. A ladder pipe operator should always be locked into the ladder with a life belt. A leg lock must never be used.

unit does not have intercom equipment, the operators should use the set of hand signals developed as part of the training program. Regardless of how they communicate, the turntable operator must watch the pipe operator constantly and quickly move the pipe operator out of any position that becomes dangerous.

When halyards are used for vertical nozzle control, the fire fighter operating them should be positioned on the turntable. This position facilitates communication with the turntable operator. In addition, the halyard operator will be protected from injury if the ladder should contact electric wires. If the fire fighter is on the ground, engine noise could drown out his or her voice, and contact with electric wires could cause injury or death by shock.

AERIAL STREAMS FOR FIRE ATTACK

To knock down a fire, aerial streams must reach into the seat of the fire. Because of its construction, the aerial pipe cannot be advanced into the fire building; streams must be directed from outside. To make sure that the streams do reach the fire, truck crews must use the proper nozzles and place the streams properly for the fire situation and weather conditions.

Nozzles

As noted earlier, both fog nozzles and solid stream nozzles should be carried on the apparatus for aerial pipe use. Fog streams are the more effective of the two in fire attack *if* they can reach the seat of the fire. However, solid streams have a longer reach and so can penetrate further into a building and through to the seat of a fire. Thus, if the aerial unit can be placed so that a fog stream will reach into the fire, the aerial pipe should be fitted with a fog nozzle; otherwise, use a solid stream nozzle.

Both fog and solid stream nozzles are rated according to their water flow rates in gallons per minute. Either type will lose effectiveness if it is not supplied with at least its rated flow. For example, a 500-gpm nozzle supplied with 500 gpm will be more effective than a 1000-gpm nozzle not receiving sufficient water. For this reason, the aerial pipe should be fitted with the proper size as well as the proper type of nozzle. This is why a range of sizes of fog and solid stream nozzles should be carried on the apparatus.

Stream Placement

To direct a fog stream into a fire building through a window, place the nozzle at the approximate center of the window opening and set at a 30-degree angle. Aim it first at the upper part of the room, where the concentration of heat is greatest, and then sweep downward. Repeat this action as necessary to control the fire.

Place a solid stream nozzle so the stream enters the window at an upward angle. This will allow the stream to strike the ceiling, break up, and spread water over a wide area (Figure 11.7). For maximum penetration, theoretically the stream should enter just over the window sill at a fairly small angle. However, the stream might then hit the building contents rather than the ceiling. A somewhat larger entry angle provides good penetration and will probably allow the stream to clear interior obstacles. Too large an entry angle will lose penetration, and the water will cascade down just inside the window.

Direct a solid stream straight into a window when maximum penetration is the most important consideration. However, never direct it down toward the

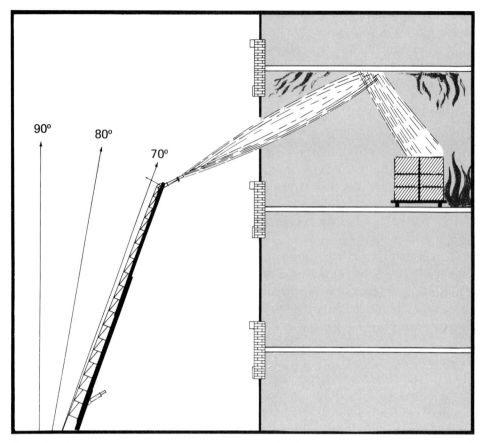

Figure 11.7. For maximum effectiveness, fog and solid stream nozzles must be placed and operated properly.

floor from a position above the window; this will have no effect on the fire and will only add much weight to a structure probably already weakened by the fire. (A water flow rate of 500 gpm adds about 2 tons of water to the building in 1 minute.)

To be most effective, an aerial pipe stream (either fog or solid) should be moved horizontally back and forth across the fire area. Also move it up and down, for maximum coverage. The amount of movement depends on the extent of the fire, but normally solid streams should be moved across the full width and height of the window. Although nothing can burn directly under a stationary stream, the fire can spread away from it, so movement of the stream ensures coverage of a good-sized area in and around the fire.

In heavy smoke, it may be difficult to determine whether or not the stream is entering the building. Fire fighters should look for steam and white smoke as indications that the stream is penetrating the fire area. In the absence of these signs, and if it is possible to do so safely, officers should visually check close to the building. If a visual check is not feasible, they should listen for the sound of the stream hitting the building and should look for heavy water runoff. Either of these signs indicates that the stream is not entering the building. The stream should then be adjusted until both indications have disappeared and steam can be seen billowing from the structure.

Wind and Thermal-Updraft Effects

Sometimes an aerial stream is adversely affected by the thermal updraft created by a large free-burning fire (Figure 11.8A). It might also be affected by wind blowing across the stream (Figure 11.8B) or toward the aerial unit. Either the updraft or the wind will tend to break up the stream, reduce its reach, and thus render it ineffective.

Fog streams, made up of smaller (and lighter) water droplets, are affected to a greater degree than solid streams. The wider the fog pattern, the more it will be affected. If a fog stream is being broken up by winds or thermal updrafts, the nozzle can be adjusted to a narrower pattern (30 degrees is the most effective) or to the so-called "solid stream" position. However, a solid stream from a fog nozzle is not as effective as one from a solid stream tip.

Move the nozzle very close to the fire building — inside a window if possible (Figure 11.8C). If the fire situation prevents this, shut down the pipe and replace the fog nozzle with as large a solid stream nozzle as can be adequately supplied with water. The solid stream will hold together much better than the fog stream, and a heavier stream will hold better than a lighter one.

Weakened Structures

At most structural fires where aerial streams must be used, the streams will be directed into the structure from positions close to it. If the structure shows signs of having been weakened by the fire, or chimneys, roof-mounted billboards, or other features seem ready to collapse, the aerial unit must be moved back away from the building. This will put the aerial pipe in a less desirable position, but is necessary to protect the equipment and fire fighters.

When a fog stream is being used to fight the fire, check it after the aerial pipe is moved. If the stream cannot reach into the fire from the new position, replace the fog nozzle with a solid stream nozzle, which also should be checked. If the solid stream is helping control the fire, operate it until the fire is knocked down. If not, it should be shut down, since it will only be adding water to an already weakened structure.

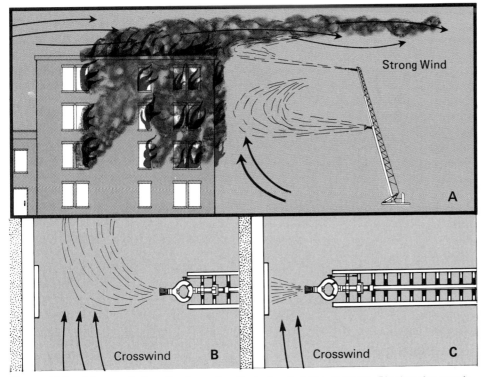

Figure 11.8. Strong winds will adversely affect elevated fog streams. Placing the nozzle close to or into the window will alleviate the problem.

Shutdown

Aerial streams should be used only as long as fire, steam or white smoke is visible in the area covered by the stream. The steam and white smoke indicate that the stream is hitting the fire. When they are no longer visible, the fire has apparently been put out in that area, and the streams should be shut down. Continued operation would only add to the water load in the building and the strain on the water supply system.

Improper Use of Streams

An incorrectly used aerial stream can cause unnecessary property loss and can result in injury to fire fighters. The two most common errors are directing streams through roof holes and directing them toward fire fighters advancing interior handlines.

Roof holes. Aerial streams should not be directed into a hole burned through the roof or opened for venting. The streams will destroy the venting action and drive heat, smoke and gases back into and through the building (Figure 11.9). Even when fire shows through the roof along with vented combustion products, aerial streams should not be directed into the roof hole.

To protect a roof from ignition, direct an aerial stream onto the roof *near* the opening. This will allow the water to flow around the opening and down along the edges of the hole. The roof will be protected, and the combustion products will still be able to escape.

When a roof (or a good portion of it) collapses, it could be that only aerial streams are able to control the fire in the area of the collapse. An aerial stream can be used in such a situation because, usually, the roof hole will be large enough so the stream does not interfere with the venting action.

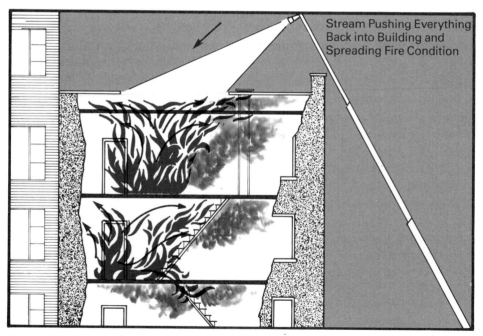

Figure 11.9. Elevated streams must not be directed into holes burned through the roof or made in the roof for ventilation.

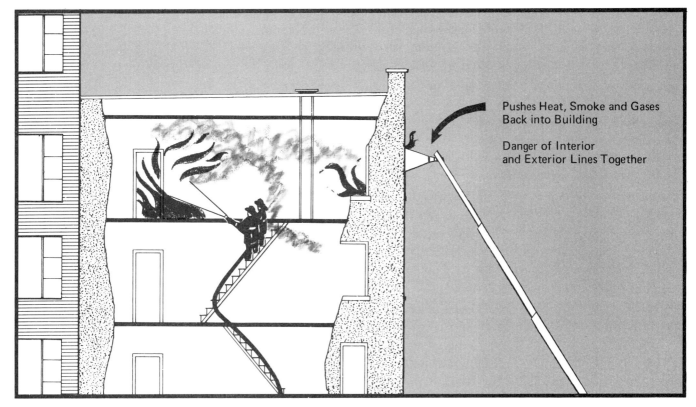

Figure 11.10. Elevated streams should not be directed into areas where fire fighters are advancing handlines.

Interior handlines. Aerial streams generally should not be directed into a building area in which crews are operating with handlines. Aerial streams may endanger fire fighters on the interior lines by pushing heat, smoke and fire back into the building (Figure 11.10). If the aerial streams are effective, great volumes of steam could engulf those inside the building. Although proper venting will reduce the severity of these effects, it is usually safest not to use aerial streams and interior handline streams together on the same area of a fire.

However, when properly coordinated, aerial streams can be used in conjunction with handlines. For instance, the size and location of a fire may warrant the use of an aerial stream as soon as fire companies arrive at the scene. The pipe stream can be placed in service while handlines are being advanced into the building through interior stairways or by other methods. The aerial stream might be able to knock down much of the fire before the handlines can be moved into effective positions.

In some cases, handlines are advanced to positions from which they can hit fire inaccessible by the aerial stream (owing, perhaps, to the depth of the building or interior construction features). Such a combined attack could be most effective in knocking down the main body of fire. Once the fire has been knocked down or greatly reduced, the aerial pipe can be shut down and the fire extinguished with handlines alone. As always, adequate ventilation will increase the effectiveness and decrease the hazards of such operations.

AERIAL STREAMS FOR EXPOSURE COVERAGE

Exposure protection is second only to rescue as a basic fire fighting objective. The importance of the truck company in checking for fire extension — that is, in protecting interior exposures — was discussed in Chapter 6. This section concerns the protection of outside exposures for the purpose of preventing exposure fires.

Outside Exposures

The term "exposure fire" is applied to outside exposures. Such a fire spreads from one structure to another or from one independent part of a building to another part, perhaps across a court or from one building wing to another.

Spaces between buildings (or between piles of stacked material) and unpierced fire walls are the major deterrents to exposure fires. They are of great assistance to fire fighters when severe outside fires develop. Outside sprinklers and spray systems are also a great help but, unfortunately, they are a rare item in fire protection equipment except in special installations.

Exposure Hazards

Truck officers and crews, as well as engine company personnel, must be familiar with potential exposure problems — conditions that promote the exterior spread of fire — in their territory. They should also be cognizant of the factors that affect the severity of an outside exposure problem. These factors include:

- Recent weather
- Present weather, especially winds
- Spacing between the fire and the exposures
- Building construction materials and design
- Intensity and size of fire
- Location of fire
- Availability and combustibility of fuel
- Size of the fire force
- Fire fighting equipment on hand

The worst combination would be recent dry weather, strong winds blowing toward exposures, an area of closely spaced frame buildings, a severe fire that is difficult to reach, plenty of easily ignited materials located between the fire building and exposures, limited personnel and apparatus response to the first alarm, and poor water supply.

Of these factors, the fire department normally has control of only the fire force and the equipment responding to the first and additional alarms. For this reason, the first alarm assignments of truck companies should be reviewed periodically, keeping in mind the contribution that these units can make toward exposure protection.

Where the exposure hazards are great, the number of companies responding to a first alarm should be increased. It is far better for these units to be on the scene when they are not needed than to have them in the station when they are needed on the fireground. This is especially true when truck companies are located some distance from each other, as is usually the case with mutual aid companies.

Exposure Protection

Exposure fires can be caused by radiated heat, convection or a combination of both. Properly located and operated aerial streams do much to protect against this exterior fire extension. Where wind is a factor, exposures on the downwind side of the fire are in the most danger from radiated heat and from embers carried by convection and wind currents. When a choice must be made, lee-side exposures should be protected first, before exposures on the windward side are covered.

Choosing the stream. Fog streams are more effective than solid streams for exposure protection, provided they can reach the area to be protected. If they cannot, because of the distance to the exposure or because of wind or thermal-updraft effects, solid streams should be used.

In general, the greater the intensity of the fire, the heavier the aerial exposure stream needed. However, no stream will be effective with a less than adequate water supply. If water supply is a problem, a smaller, adequately supplied, stream will be more effective than a weak stream from a larger nozzle.

Directing the stream. Since water is transparent, radiant heat passes through it. Throwing a stream between the fire and the exposure will not protect the exposure; the heat will simply pass through the stream and warm the surface of the exposure to its ignition point. Instead, the stream must be directed onto the surface of the exposure in such a way that the water washes down its sides (Figure 11.11). Only then will the water absorb heat from the exposure and keep it from igniting.

If an exposed building is taller than the fire building, its most vulnerable area is above the level of the fire. That area will be subject to both radiated

Figure 11.11. Streams must be directed on the area to be protected, not thrown between the fire and the exposure.

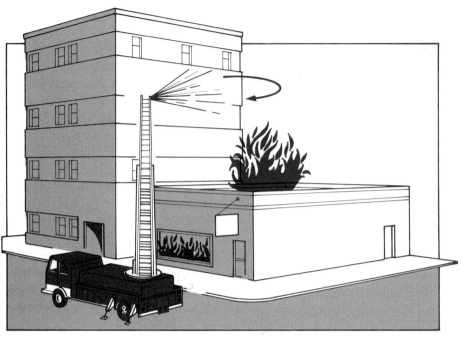

Figure 11.12. The point of greatest danger to this exposure is just above the top of the fire.

and convected heat, and also to any firebrands and embers carried by the wind (Figure 11.12). The first aerial exposure streams in service should be directed just above this most vulnerable area.

Burning embers and ignited materials can be convected up to exposures from a fire that has burned through the roof of a building or has ignited piles of material in a storage yard. A properly positioned aerial stream, directed into the fire and its smoke column, can be of great help in decreasing this exposure hazard. An exposure stream can often be used for this purpose. That is, the stream can be alternately directed onto the exposure and onto the fire. In many other situations, such action can help knock down the fire or decrease its intensity in the area of the exposure. However, the pipe operator must be careful not to give too much attention to the fire and too little to the exposure.

When an exposure is so long that one aerial stream cannot protect it completely, an attempt must be made to position a second stream for complete coverage. If an aerial pipe cannot be positioned to cover the unprotected part of the exposure, other heavy streams or handlines must be used. In many cases, a combination of aerial streams, other heavy streams, and handlines will be required for complete exposure coverage and fire control.

ELEVATED HANDLINES

An aerial pipe is a heavy stream device that can deliver up to 1100 gpm. In some situations, an elevated stream, but not a heavy stream, is required. Handlines should be operated from an aerial ladder or platform in these cases.

Operating from a Platform

The handline, with the nozzle attached, should be tied to the platform railing before the platform is raised. The nozzle and about 1 foot of hose should extend out in front of the railing; this will allow the stream to be moved in all directions.

Do not raise the platform until the nozzle operator is secured to a railing with a life belt. If the stream is to be directed in through a window, place the basket so the tied-in nozzle is at the center slightly below the bottom part of the window. A fog stream at a 30-degree angle can then be aimed into the upper part of the room; a solid stream can be aimed for maximum penetration; and the operator will be safe from fire or gases that might be pushed out the window.

Once the fire in the area of the window is knocked down, the basket can be moved above the sill to allow horizontal penetration, if necessary. When the fire is thoroughly knocked down, the handline can be advanced into the building. How this is done depends on the size of the line and the height above the ground.

With a 1½-inch or 1¾-inch line, the nozzle can be shut down and the line untied and moved into the building. Crews below can assist by pushing slack up to those in the window. A rope tied in behind the first or second coupling can be a great help in getting the charged line up. With a 2½-inch hose, use a rope to advance the line into the building quickly. Draining the line will speed up the operation.

If not enough slack is available, the line should be shut down and the coupling closest to the building disconnected. More sections of hose should be added to the line as required, and the line moved into the building and charged.

Operating from an Aerial Ladder

Usually, an elevated handline is directed into a fire building through a window. If fire is showing at the window, the aerial ladder should be raised but not placed close to the building. The fire fighter operating the nozzle should take the line up the ladder to the proper position and tie it in to the center of the ladder. The nozzle should be approximately centered in the window just below the sill. Attack and knock down the fire before the ladder is moved in toward the building. Then the fire can be hit again, if necessary.

If no fire is showing, place the ladder at the window before the fire fighter on the nozzle takes up the line.

In either case, the fire fighter working on the ladder must wear a life belt. The ladder must not be extended or retracted once the fire fighter is on it. The ladder operator must be careful not to activate the extension-retraction control when rotating or moving the ladder, as this could seriously injure any fire fighters on the ladder.

The elevated handline can be untied and moved into the building as previously described when it is safe to do so.

In some situations, a handline is operated from an aerial ladder while the ladder pipe is in operation above. This allows fire fighters to attack the fire at two levels at the same time. However, since there is a crew member on the aerial below the top of the ladder, it cannot be extended or retracted. Depending on the situation, this (and the manufacturer's loading and flow ratings) might restrict the performance of the ladder pipe.

SUMMARY

The elevated streams provided by aerial apparatus are important in attacking fires and protecting exposures. Truck crews must know how to quickly set up their aerial pipes and must cooperate with engine companies to supply them with water. Officers and aerial operators must be able to accurately size up any given fire situation and spot the turntable properly.

A nozzle of the proper size and type must be chosen on the basis of the intended use of the stream, required reach, water supply, wind conditions, and the effect the fire could have on the stream. In most cases, solid streams of the proper caliber will be most effective.

Elevated handlines can be operated from aerial units where height is required and no heavy streams are needed.

CONTROL OF UTILITIES 12

The utilities within a building — that is, heating, ventilating and air conditioning systems, electric wiring, gas, and water pipes — often cause problems during fire fighting operations. They can contribute to fire extension, add fuel to the fire, and lead to extremely hazardous conditions.

- Forced-air heating, air conditioning and air circulation systems can blow smoke, gases and fire through the air ducts to uninvolved parts of the building, creating additional fire fighting, rescue and safety problems.
- Oil and kerosene burners or their fuel supply systems can become damaged and leak fuel, thus feeding the fire.
- Gas lines can be damaged and leak gas. If this gas ignites immediately, it will feed the fire; if not, the leaking gas will collect and possibly explode.
- Electric wires that become exposed or burned in two and electric fixtures that are damaged and shorted can cause severe shock if contacted by fire fighters or their streams.
- Water flowing from broken pipes increases fire fighting hazards by adding to floor loads or by flooding lower levels where fire fighters are working. The water also can cause unnecessary damage to the structure and its contents.

To minimize or eliminate such problems, it is important to monitor and control the utilities within the fire structure. This job is usually assigned to truck company personnel.

PREFIRE PLANNING FOR UTILITY CONTROL

The type, size, placement and complexity of the utilities in a particular structure usually depend on the occupancy. In commercial and industrial occupancies, they may also depend on the activities performed within the building. Therefore, knowledge of the assigned territory is as important in utility control as it is in other truck company operations. Truck companies should become aware — through prefire inspection and planning — of any unusual utility systems within their territory. They should also have a thorough understanding of the hazards associated with each type of utility.

Operating Controls

Utilities are, for the most part, controlled through power switches and shut-off valves. The kind of controls installed in a utility system and the way they function depend on the type, size and age of the system. Fire fighters must be familiar with the various kinds of operating controls. In particular, they should be fully aware of what these controls will do when activated or deactivated.

During prefire planning, truck crews should familiarize themselves with the locations of power switches and shut-off valves, and determine whether each of these devices controls all or part of its utility system. This is especially important in the case of large, complex systems, such as those in manufacturing and processing plants, large commercial buildings and apartment houses, schools, and hospitals. Such installations should be inspected as necessary.

In smaller structures, including one- and two-family dwellings, garden apartments, stores, garages, and small commercial buildings, neighborhood building patterns sometimes result in a common location for each utility control in each type of occupancy. Whether or not this is the case, truck company personnel should have a good idea where these controls will be found, when needed.

Building Codes

National and local building codes can be helpful in determining the locations of operating controls. Sometimes these codes require that the controls be placed in designated positions which depend on the type, size and layout of the building. Officers and fire fighters should be aware of the effect of these requirements on the structures — both large and small — in their territory.

FORCED-AIR SYSTEMS

In a forced-air system, treated air is pushed through a series of ducts to the living and working areas of a structure. At the same time, air is drawn out of these areas through a second set of ducts. This "return air" is then treated and forced back to the living and working areas. The air can be treated by heating, cooling, filtering, dehumidifying, or a combination of these processes. It is forced through the ductwork by one or more fans or blowers (Figure 12.1).

In case of fire, the two sets of ducts form a system of channels for the horizontal and vertical spread of heat, smoke, gases and flame. Dust and fibrous materials tend to collect along the bottom of the ducts, building up a fairly thick mat of combustible material over a period of years. If this material is ignited or heated sufficiently, the fire might flash through the entire ductwork system.

Forced-Air Blower

Action of the fan or blower compounds the problem in three ways. First, it increases the efficiency of the ducts as channels for fire spread. Second, it

Figure 12.1. Air return draws in smoke, heat and gases, spreading them to all rooms.

draws fire and heated gases into the ductwork through the return-air inlets, perhaps involving the ductwork. Third, it forces hot combustion products from involved areas to uninvolved areas through the treated-air ducts. If these hot combustion products are discharged onto combustible material, it will ignite. The fan will then repeat this cycle, forcing the heat from newly involved areas throughout the building.

For this reason, it is imperative to check the blower or blowers in an involved structure, and to shut them down if they are operating. Usually, it is best to simply shut down the entire heating or cooling system.

Some of the larger forced-air systems contain smoke detectors which monitor the air in the ductwork. If a detector senses smoke, it automatically shuts down the system. Fire fighters should be familiar with such systems, and with the locations of their manual control switches. This type of system must also be checked to make sure it has a shut down; if not, it must be shut down manually.

Dampers

Some larger systems also are equipped with dampers which close automatically to seal off portions of the ductwork and isolate the fire. The damper is spring loaded and held open by a fusible link. When the heat of a fire melts the link, the spring pulls the damper shut.

Automatic dampers are usually located where ducts pierce fire walls or other major dividing walls. An easily opened inspection plate is located near each damper. The plate allows a quick check to determine whether the damper has closed. If a closed damper has been subjected to heat or fire, the duct on both sides of the damper must be inspected for fire spread.

These dampers are effective in stopping the spread of fire. However, experience has shown that large amounts of smoke and gases can travel past the dampers and spread through the building before the fusible links become hot enough to melt. This is why some large systems have dampers actuated by smoke detectors rather than by fusible links. Smaller systems, such as those in dwellings, do not normally have dampers. In any case, the forced-air system must be shut down at a working fire.

Heating Systems

The forced-air heating systems usually found in one- and two-family dwellings are designed so that a burner (oil, kerosene or gas) produces the heat. The heat warms the air in a plenum located above the burner. A blower forces the warmed air through the ducts to the living areas and draws room air back to the plenum. The burner and blower are controlled by separate thermostats. A thermostat in the living area turns the burner on to warm the dwelling, and off when the desired temperature is reached. A thermostat in the warm-air plenum operates the blower as long as the air in the plenum is above a set temperature.

When there is a fire in the living area, the increased temperature causes the burner thermostat to shut down the burner. However, the hot return air increases the plenum temperature, allowing the blower to continue to operate. As noted above, this action aids in the spread of fire, so the system must be shut down.

Larger, more complex forced-air heating systems work on the same basic principle. Some, however, are divided into separate sections, or zones, each with its own blower. In larger structures, entire ceiling areas are sometimes used as part of the air-return system. In such cases, fire drawn into the system is not confined within a duct, but is easily spread over a large, unprotected area (Figure 12.2).

In medium- and high-rise buildings, a single duct "circuit" may serve several floors. Smoke, heat and gases from a fire on one floor are easily carried to other floors, endangering people who would otherwise be relatively safe. When this situation causes panic, it compounds the rescue problem.

Cooling Systems

The thermostat of a cooling (air conditioning) system keeps the system operating as long as the room temperature is above the set temperature. The heat from a fire will cause the system to run continuously, in an attempt to reduce the room temperature to satisfy the thermostat setting. As with a heating system, the operating blower can easily draw fire in through the return-air inlets and spread combustion products through the ductwork or ceiling area to the area it serves (Figure 12.2).

When the outdoor temperature requires neither heating nor cooling indoors, forced-air blowers might be operating merely to circulate air through a structure. Again, the operating system must be shut down by fire fighters when a fire occurs.

Air Circulation Systems

In older buildings, including theaters, halls, and some other large structures, an obsolete type of air circulation system might still be in use. This system has large ducts that extend throughout the building. The ducts are connected to a huge air intake and a blower, usually located in the basement. In some cases, the system includes a primitive cooling setup in which cool water is circulated through large radiator-like units placed between the blower and the air intake. These units cool the air as it is blown into the system. Usually these systems vent through the regular ventilators in the building. In some cases, however, exhaust fans are used to provide faster air

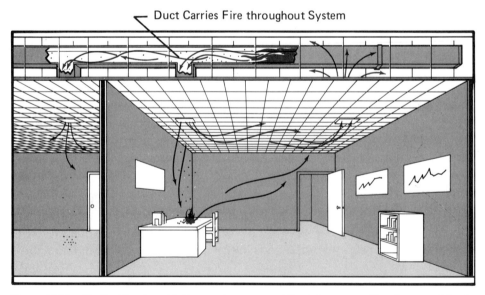

Figure 12.2. Radiant heat and falling embers spread fire into rooms. Air returns pull heat, smoke and fire into hidden ceiling spaces.

movement in hot weather. These venting fans are different from the venting systems installed over stages. The latter are used to vent the stage area and deter the spread of fire into the auditorium.

These air circulation systems are almost always controlled manually rather than automatically. Therefore, they must be shut down manually, through a master power switch, when a working fire is encountered.

The problems that can be created by forced-air systems and the extreme importance of shutting down these systems when a working fire is encountered have been discussed. It should be noted that some forced-air systems are designed so they also can be used to exhaust smoke from the building *after the fire has been knocked down.* Truck companies should determine, through prefire inspection and planning, where such systems exist.

HEATING UNITS AND FUELS

When a heating unit is involved with fire or in danger of becoming involved, both the unit and its fuel supply must be shut down. Heating fuels are, by their nature, combustible; it is important to make sure that they do not add to the fire fighting problem.

Oil Burners

Oil burners are used to supply heat for forced-air, hot water, or steam heating systems. The fuel oil is stored in a tank and pumped to a burner in the furnace. In the burner, the oil is vaporized, mixed with air, and burned. The burner, fuel pump, and heat circulating device (blower or water pump) are operated electrically.

Small systems. Small oil burners, such as those found in one- and two-family dwellings, burn No. 2 fuel oil, which is similar to kerosene or diesel fuel. The storage tank can be located in the basement or above ground, against the outside wall that is closest to the furnace. A fuel line (usually copper) carries fuel from the tank to the burner.

Most building codes require that an emergency power shut-off switch be located at the top of the basement stairs (if the unit is in the basement), at an outside basement entrance, or just inside an entrance to the building or the basement. The emergency switch is bright red. When it is turned off, it cuts off all power to the unit, including the fuel pump, and thus it will ordinarily stop the flow of fuel.

However, if there is a break in the fuel line, and the fuel tank is above the level of the burner, fuel will flow down toward the burner. The flow of fuel can then be stopped by closing the fuel shut-off valve located on the supply line, right at the fuel tank. A supply line made of copper tubing can be squeezed together with pliers to stop the flow of fuel if the valve will not work. If the supply line is made of heavy pipe, it will be necessary to cut the pipe and plug it.

Large systems. Large oil heating units, such as those in apartment houses, office buildings, factories, schools and hospitals, burn a heavy oil, usually No. 6 fuel oil. Since the heavy oils do not ignite as readily as No. 2 oil, many of these systems include a device that preheats the fuel oil before it is burned.

The fuel is stored in underground tanks or in large aboveground tanks, usually located some distance from the building. The oil is pumped from the tanks to the preheaters and then to the burners. The emergency power shut-off switches shut down the pumps, preheaters and burners; this usually stops the flow of oil from the storage tanks. However, on occasion, a siphon is set up within the fuel line and fuel continues to flow even though the pumps have stopped. Aboveground and elevated tanks are especially prone to siphoning, so the fuel flow should be checked after the power is turned off. If possible, the manual fuel shut-off valves should also be closed.

Some large oil burning systems are sectionalized, with a separate subsystem serving each part of the structure. In such cases, it might be necessary to shut down only one subsystem to gain control of the fire. However, if there is any doubt, the entire system should be shut down. Building engineers can be of help in determining the proper course of action.

Kerosene Heaters and Stoves

Kerosene is still used as a fuel for both heating and cooking, but to a lesser extent than oil or gas. Kerosene heaters and cook stoves are found in rural areas and in poorer urban areas, where the cost of gas or oil is prohibitive or older central heating systems have deteriorated.

Kerosene heaters are self-contained units — that is, they are not connected into a separate heat distribution system as are oil burners. Some are designed to stand on the floor of a room, providing heat by convection out of the top and by a fan that blows heat out at floor level. Other kerosene heaters, known as floor furnaces, can be placed below the floor. A grate in the flooring above the unit allows the heat to rise into the living area by convection.

Kerosene cook stoves resemble standard gas or electric kitchen ranges. Both stoves and heaters can contain fuel tanks or can be fed fuel from an outdoor storage tank. Some are designed to operate from either type of fuel supply.

Kerosene-burning units do not usually have remote emergency switches; all the controls are on the unit itself. This makes it necessary to cut off the electricity in the building to keep fans from operating if the unit cannot be reached during an emergency. In addition, the fuel shut-off valve at the outdoor tank must be closed to keep fuel from continuing to flow. Otherwise, the kerosene will feed the fire if the fuel supply line is damaged.

When fire has involved the area around a heater that has a self-contained fuel tank, the tank might explode and throw flaming fuel over the entire room. When it is known that such heaters are in use, fire fighters should approach involved rooms with extreme caution, and with hoselines ready. If the tank has already exploded, standard fire fighting operations should be used to bring the situation under control.

Gas Units

Gas, either piped (city gas) or bottled, is used for heating and cooking and in some manufacturing processes. City gas is piped through meters to the point of use; it is usually a natural fuel but could be manufactured. Bottled gas is a liquefied petroleum product (LPG), such as propane, stored in heavy outdoor tanks near the point of use. Both piped and bottled gas are highly combustible; a gas supply or outlet threatened by fire should be closed off to prevent ignition or explosion.

City gas. Some cities require that an emergency valve (gas shut-off) be located in plain sight outside the front of any building served by gas lines. The

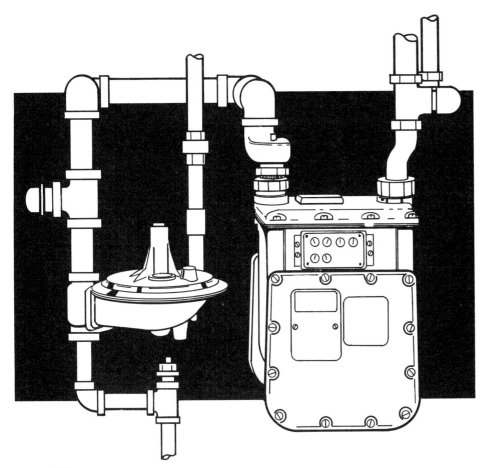

Figure 12.3.

shut-off is usually painted a distinctive color, and marked "emergency gas shut-off." Outside main gas valves might also be located at or near outside meters, or just outside the locations of indoor meters. When an outside main valve is not provided, the gas shut-off is located on or near the gas meter, inside the building. Unfortunately, this is the case when the meter is located in the basement. When a basement fire or a serious first floor fire makes it impossible to reach the meter, the street valve must be used to stop the flow of gas.

The local gas company is sometimes of help in locating the outside shut-off or street valve. However, the degree of cooperation depends on the individual gas company. Many fire departments have had difficulty obtaining immediate assistance from gas companies and, as a result, have encountered serious fire problems.

Gas meters. The gas meter is usually the weakest link in the gas supply system. A meter located in an involved area will generally fail before the piping fails. When the meter fails, the escaping gas will ignite and form a burning jet. The escaping gas should be allowed to burn while the immediate area and exposures are protected. No attempt should be made to extinguish the gas with streams. Flames at a meter — or at a damaged gas line — should be extinguished only by shutting off the flow of gas (Figure 12.3).

If a gas fire is extinguished before the supply is shut off, the gas will probably reignite. Worse, it might collect and eventually explode, causing further damage to the gas supply system or the building, and perhaps injury to occupants and fire fighters.

In many apartment buildings served by city gas, a separate meter is provided for each unit. The meters are usually located in a central bank in the basement of the building. This arrangement allows fire fighters to shut off the gas selectively, instead of shutting off the flow to the entire building. More important, once the fire is extinguished, it allows undamaged gas appliances to be put back into operation while the supply of gas is held back from apartments with damaged appliances or supply lines. Restoration of service should be done only by gas company personnel.

When a fire involves the area in which a bank of gas meters is located, large amounts of gas can be released. The gas will add to the intensity of the fire or might collect and explode. The street valve should be closed under such conditions.

Truck company personnel should be familiar with the locations of inside and outside shut-off valves and street valves. The apparatus should be equipped with any special tools or "keys" needed to close street valves or the large valves used in industrial or large commercial installations.

Bottled gas. Bottled gas for home use is usually stored in one or more outdoor tanks located near or directly against the house. A copper line carries the fuel into the house to the appliances. Usually there are no meters, but a gauge and shut-off valve are located at the top of the tank (or on the supply line if several tanks are being used). Closing the valve will shut off the flow of gas into the building. If for some reason the valve will not operate or will not completely shut off the flow, the copper line should be squeezed shut with a pair of pliers (Figure 12.4).

If a fire in the building endangers the tanks, they and the exposed supply line should be cooled with a stream. Each tank has a relief valve or plug to relieve the tank pressure in case of fire. If this valve has been actuated, a jet of burning gas might issue from the tank. The gas should be allowed to burn while the tank and the area around it are cooled with a fog stream. Extinguishing the flame would not stop the flow of gas. The unburned gas might collect and reignite, causing a flash fire over a large area, or an explosion.

The gas used in large industrial plants is stored in tanks located above or below ground level. In most installations, buried piping carries the fuel to the plant. Shut-offs can be located at the tanks or on the supply line outside the building. In addition, sectional valves might serve various parts of the plant, and a bank of remote control valves might be located at an engineer's station in the plant.

It is important that truck company personnel be familiar with any such installations in their territory. Truck crews must be able to locate and close off the appropriate valves quickly when the fire situation requires such action.

ELECTRIC SERVICE

When a working fire is encountered, officers must consider shutting down the electric service in the involved area or in the entire structure. Fire extending into walls, ceilings or floors could have burned the insulation off electric wires; electric fixtures might have been damaged and their wiring exposed. Fire fighters who contact the bare wiring are subject to serious injury or death. Streams can hit exposed electric features, which will cause electric shock. Truck crews might strike exposed wiring with their tools when checking for fire extension, ventilating, or performing other duties.

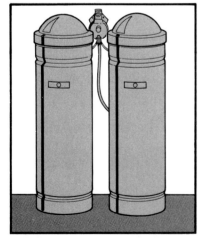

Figure 12.4.

The best way to prevent injury from electric shock is to shut down the electric service in the fire building. However, in each situation the need for electricity during fire fighting operations must be balanced against the dangers it presents. For example, suppose fire companies arrive at a working apartment house fire at night, and find occupants attempting to leave the building. It would be senseless to immediately cut off the electric supply and, with it, the lights being used by escaping occupants. The confusion and panic generated by the sudden darkness would far outweigh the benefits in terms of fireground safety.

The electric service in a fire building can be used by fire fighters to power portable lights, fans and electrically operated tools. The building lights can be of help in search, fire attack, and other operations. However, when the condition of the building indicates that electric features might endanger fire fighters, the electricity should be shut off in the fire area or, if necessary, in the entire building. In the example above, the officer in charge might have the electricity shut off once it was ascertained that all occupants had escaped the fire building.

Main Power Switches

The electric service provided to a building depends on the type of occupancy. Dwellings usually require 110-volt or 220-volt electricity for lighting, heating and electrical appliances. Industrial plants often require up to several thousand volts for manufacturing processes. The number, type and locations of power switches depend on the type of service provided.

Dwellings. The modern electric service to a one- or two-family dwelling can be shut off completely by pulling (removing) the electric meter. Service to particular areas of the house can be shut off at the fuse box or circuit breaker box.

When the power switches are located in the fire area and cannot be reached, it might be necessary to cut the electric service line to the building. If possible, this should be done by power company personnel. However, sometimes — as when power company personnel are delayed — the fire department must do the job. Many departments equip their truck companies with "hot sticks" for this purpose. A hot stick is an insulated wire cutter mounted on a long pole, with a cord that operates the cutter when it is held overhead. These devices are available commercially; the manufacturer's performance rating should be checked before any such tool is purchased.

Commercial and industrial structures. The situation is similar at larger buildings, although the voltages may be much higher. Usually, a main power switch and several sectional switches allow the power to be cut either completely or selectively. Especially at night, it could be important to shut off the electricity in the fire area only, so that the lights of nearby areas can be kept on during fire department operations.

In most commercial and industrial structures, electric panels and lines that carry unusually high voltages are so marked. The presence of heavy, uninsulated busbars indicates the use of high voltages in an area. Fire fighters might also find large cartridge-type fuses; a fuse puller is required to remove these fuses safely. Where possible, fire fighters should seek the assistance of building engineers or maintenance personnel in pulling fuses or shutting down a high-voltage electric system.

Elevators

Elevators are usually powered by their own electric circuit, separate from that for the remainder of the building. This allows the elevators to be operated even when the electricity in the building itself is shut off. Therefore, the elevators can be used by fire fighters for rescue and fire control, conditions permitting.

The main power switch for the elevators is located near them, usually in the basement. The power to the elevators should only have to be cut off when the fire has control of the elevator shafts or the motor house, usually located on the roof.

WATER PIPES

Although a damaged water system will not contribute to fire spread, it can add to operating problems or become hazardous to fire fighters. An uncontrolled flow of water will cause unnecessary damage to the building and its contents and add substantially to the floor load. The water could also flood lower levels where fire fighters are working.

In cold climates, there is the added danger that the water will freeze, causing additional structural damage and difficult footing for fire fighters. Water freezing in water or drain pipes can burst the pipes, and then be released when the temperature rises above the freezing point.

Further, an accumulation of water might become charged with electricity if the power has not been shut off. Fire fighters cannot depend on their boots for protection against electric shock. The large amounts of carbon added to the rubber used in manufacturing boots will conduct electricity to fire fighters' feet. For these reasons, truck companies should maintain control of both the water system and the power system. A damaged water system should be shut down as soon as possible, to minimize problems due to excessive water or freezing and to reduce the extent of required salvage operations.

Water Shut-off Valves

In communities that meter water, a street shut-off valve is usually located near the meter at the front of each building. There may be a shut-off valve in the street even where city water is not metered. In most communities, there is a pattern to the placement of these main water valves; truck companies should be aware of both the pattern and the exceptions. Special keys might be needed to close street valves.

In buildings with basements, the main shut-off valve is often located on the incoming water line just inside the basement wall. In buildings without basements, the shut-off valve should be located on the interior of an outside wall, in some sort of utility area. Normally, there are separate shut-off valves at sinks, basins, toilets and individual water-using units in industrial plants.

In large structures, the water system often is divided into sections for water control during repairs; the locations of such sectional controls should be determined as part of prefire planning.

Boilers and Heating Units

When it is necessary to cut off the flow of water to industrial boilers or to water or steam heating units, the heat must also be cut off. Otherwise, a dangerous situation could be created. Water service should be restored to such units only as recommended by the manufacturer.

SUMMARY

It is evident that truck crews assigned to utility control at fires must have a thorough knowledge of the various utilities they are likely to encounter. Forced-air systems; oil, kerosene and gas burners; electricity; and water supply systems should be monitored and controlled to assist in fire control operations and to keep injury and damage to a minimum.

Truck company personnel should know how to locate and operate the types of power switches and shut-off valves used in their territory. They should know which structures contain sectionalized systems and be aware of what parts of the structure are served by each section. They should not hesitate to shut down all or part of a utility system when such action will result in safer or more efficient overall fireground operations.

OVERHAUL 13

When the main body of fire is apparently extinguished, overhaul operations begin. Overhaul is the hard, dirty job of searching for the remnants of a fire — sparks, embers, small concealed fires — to ensure that the fire is completely out. The main purpose of overhaul is to make certain that no trace of fire remains to rekindle after the fire force has left. A second purpose is to leave the structure in as safe a condition as possible, especially if it is to be partially or completely occupied soon after the fire. Because water damage can occur during overhaul, fire fighters must protect undamaged goods and furnishings as part of these operations.

Cleaning up the premises, although perhaps good for public relations, is not a necessary part of overhaul. Debris that could be dangerous to occupants or which must be examined during overhaul should be removed from the building. However, truck companies should not be kept out of service for cleanup, especially in a structure with its own maintenance force.

At many working fires, overhaul is the toughest fire fighting assignment. It requires knowledge of fire travel and building construction, expertise in the use of overhaul tools, and the stamina and muscle for prolonged periods of hard work. Unfortunately, in many cases fire fighters already exhausted from strenuous fire fighting operations are immediately assigned to overhaul work.

Tired crews sometimes try to work too quickly and tend to take chances in an effort to get the job finished. This often results in mistakes and in injuries to fire fighters. During fire attack and related operations, truck forces must work quickly and take calculated risks as required, but overhaul begins after the emergency is over so there is no reason to rush or take chances. There is time enough for officers to look over the situation and develop an orderly and safe overhaul plan. Yet, in spite of this, the injury rate for overhaul operations is relatively high.

Two procedures can help reduce the frequency of injury during overhaul: (1) preinspection by officers and (2) assignment of fresher personnel (if available) to overhaul duties, along with proper control of these personnel.

These procedures are discussed in the first two sections of this chapter; the remaining section deals with the overhaul operation itself.

PREINSPECTION

Before any fire fighters are sent into a building for overhaul operations, the fire area must be checked thoroughly. The building damaged by the fire might also be strained or damaged by the weight of the water used during fire attack and extinguishment. There might be holes in the floors and roof; stairways might be hazardous; a portion of the building might be unsafe to enter; and other dangerous conditions could exist. The groups of fire fighters who will be overhauling the building may know little about such unsafe areas or their locations. Unless the building is inspected and unsafe areas marked or rendered safe, there are bound to be accidents and injuries, whether or not a fire fighter who worked the area will guide each group.

The extent of this preinspection, as well as the extent of the entire overhaul operation, will depend on the size of the fire. If the fire were small, perhaps confined to one room and its contents, an inspection of the fire room and the rooms around it will probably suffice. If a large part of the building were involved, the entire overhaul area should be cleared of fire fighters, and the appropriate officers should inspect it. In some cases, the entire building will have to be examined.

The purpose of the inspection is to make sure that the area in which overhaul operations will be conducted is safe. The inspecting officers should see to it that holes in the flooring are safely covered or barricaded, that unsafe stairways are marked or roped off, that structurally unsound areas of the building are marked in an obvious way, and that portable lights are placed in dark and dangerous locations (Figure 13.1).

When damage has been extensive at a night fire and the entire area cannot be properly lighted, the overhaul operation should be delayed until daylight. The combination of extensive damage and poor visibility is extremely conducive to accidents. Watch lines should be established to extinguish any fire that might be rekindled, and some truck crews should remain on the scene to assist engine companies. Other units should be returned to service until it is safe to begin overhaul.

PERSONNEL

If it is safe to begin overhaul operations soon after the fire is knocked down, fire fighters in the best physical condition should be assigned to these duties. Those who have been fighting a fire for some time will be tired and worn down by heat, smoke, and the weight of their air masks. Truck crews who have been moving constantly from one assignment to another are generally not in shape to conduct a careful examination of the premises. They deserve a rest before reassignment, and the beginning of an extensive overhaul operation is a good time to give them one. If they are needed later, they will be more alert after their rest and better able to perform assigned duties (Figure 13.2).

Overhaul operations should be assigned to fresher truck company members who were not directly engaged in fire fighting operations. Those who may have been assigned to check exposures, place ladders, or assist with ladder-pipe operations can be used. Crews arriving on their own (off-duty paid personnel or late-arriving volunteers) are candidates for overhaul duties. If necessary, fresh truck crews should be called to the fire scene for overhaul.

Figure 13.1. Before overhaul operations begin, the area must be inspected and necessary safeguards must be put in place.

Control of Personnel Movements

Fire fighters who have not been inside the fire building during the fire fighting operations will not know which areas have been damaged or where the structure might be weak. The outside appearance of the building may give them little information about interior conditions. It is therefore extremely important that such personnel be controlled and guided within the structure.

No one should be permitted to enter the fire building without first reporting to an officer. Fire fighters entering the building without reporting may find themselves in serious trouble if no one knows that they are inside. Nobody should enter the building if not wearing protective clothing. No one should be allowed to enter alone (Figure 13.3).

Overhaul operations should be assigned to truck companies as a whole or to members formed into groups. These companies or groups, and not individuals, should be assigned areas in which to operate. If possible, each group should have a guide who is familiar with the situation inside the building. The groups should maintain contact with each other and with the officers in charge of the overhaul operation.

The reason for these recommendations should be obvious: although the fire might have been extinguished, there is a high probability that parts of the building are unsafe. Following these rules minimizes the chance of an accident, minimizes the likelihood of serious injury if there is an accident, and makes sure that help is available if needed.

Figure 13.2. Crews who have worked during the initial attack phase should be allowed time to rest. The fresh overhaul crew should be given work assignments prior to entering the building.

Work Assignments

The basic duty of truck crews in the overhaul operation is to find existing embers, sparks and fire. This often requires opening various building features for examination or further opening spaces that were opened earlier to check for fire extension. Engine company personnel advance and handle the lines needed to extinguish any embers or small fires that truck company personnel may find.

Truck crews should be assigned to the areas above, below and around the fire area, to check for fire spread that has not yet been detected. Engine crews should be called to these areas only if a hot spot is discovered. Jobs that are neither truck work nor engine work should be shared. For example, if heavy items must be removed, both engine and truck crews should be expected to lend a hand.

PROCEDURE

The basic purpose of overhaul is to make sure the fire is completely out everywhere. This means that any area that could possibly have been in contact with fire or intense heat should be checked, whether the building is fire resistant or not. Often, the building materials resist fire, but holes, shafts and other openings that are left during construction provide perfect channels for fire spread. Truck crews should be assigned, with their tools, to check — and to open if necessary — such building features, as well as the more usual channels.

The first step in the operation is to determine the stability of the building, especially when considerable damage has been done. The building's age, type, construction, occupancy, and other features may affect its stability.

Figure 13.4 shows a dwelling in which a serious working fire was encountered. The fire is not completely extinguished; therefore, overhaul must be complete and thorough. Much water was used to extinguish the fire; during overhaul, wet masonry steps compound the danger of slipping and falling on debris and broken window glass. The large amounts of water also make the electric wiring system extremely dangerous to overhaul personnel. Broken glass lies inside and outside the dwelling. Jagged edges of broken glass have not been completely cleaned out of the window frames.

Openings have been made in the roof. The fire has weakened the roof covering as well as the roof support, and portions of the weakened roof could fall to the interior or exterior and cause injuries. The flooring has been weakened by the fire and by fire fighting operations. Weak areas must be protected from total or partial collapse.

Loosened or weakened ceilings must be removed, or they might fall and injure occupants returning to the involved structure. Interior stairways are often weakened by a working fire; they should be thoroughly investigated to determine their condition, whether or not they appear safe. Other dangerous items include hanging or fallen debris with exposed nails, storm windows, awnings, air conditioners, hanging wires, distorted and protruding pipes, and construction metal such as flashing.

Overhaul operations usually begin close to the areas in which fire fighting operations ended, since the hoselines are already there. It is important, however, that other areas also be checked at the same time. Engine companies should stretch additional lines as required, so that lines need not be moved about excessively.

Figure 13.3. Fire fighters without protective clothing should *not* be allowed into the building. Crews arriving on the scene during overhaul must report to a certain point or officer. No one should just wander into the fire building.

DANGER POINTS

Doorways... Windows... Weakened Roof, Flooring or Ceilings... Openings in Flooring or Roof... Weakened Stariways... Debris... Exposed Nails... Metals... Hanging Wires... Pipes... Electrical Wiring System... Leaking Gas

Figure 13.4. Overhaul is dangerous work, especially if the building has been damaged excessively.

Indications of Rekindling

Truck companies assigned to overhaul duties should look for flames, smoke, heat, a stronger than normal odor, and areas that obviously have been touched by flames. They should also look for vertical black streaks near

Figure 13.5. During overhaul, beginning close to the area where fire fighting operations ended, it is essential to look for signs of hidden fire.

baseboards, and blistering and discoloration on walls that have not yet been checked (Figure 13.5). Concealed horizontal and vertical spaces should be checked, whether or not they were opened during fire attack and exposure protection operations. Portable lights are of great help in examining concealed spaces or any areas that may have to be opened up. They also show up smoke that otherwise might not be seen.

An area in which undamaged cobwebs are found has probably been missed by the fire, since cobwebs tend to shrivel up in temperatures that are higher than normal.

A ceiling, floor, wall or shaft that is opened for examination often shows some sign of fire damage, usually in the form of blackened or charred surfaces. It should then be opened further, until the full extent of the fire damage is indicated, usually by a clean area that was not touched by the fire. If flames, embers or smoke show when any space is opened, the area should be wet down and then further opened until the full extent of the fire is visible.

Areas of Possible Rekindling

Walls and ceilings. If walls or ceilings have been in contact with fire and heat, they must be opened and checked for signs of fire (Figure 13.6). If they

Figure 13.6. During overhaul, ceilings must be opened until a clean area is found.

have been partially opened in the course of fire fighting operations, they should be opened further until the full extent of the fire is found. Engine crews should be called to wet down any suspicious spots or areas with a light stream, and truck crews should check adjoining ceilings and walls.

It is important that ceiling spaces be thoroughly examined, because any fire there will be guided to wall spaces and then up through the building. Ceiling spaces should be checked with extra care, so damage will be minimized. If possible, drape stock and furnishings in the area with salvage covers when ceilings are opened. At this stage, every effort should be made to protect the building contents from further damage.

A check should be made to determine whether sparks have been carried up into interior walls or partitions. If so, there is often smoke and a strong odor. Where there are such signs, an adequate opening should be made and a light stream directed up into the area to wet it down. Fire fighters will be able to detect signs of steam if a hot spot is encountered. These spots must be checked again before overhaul is completed (Figure 13.7).

Above the fire. Above the fire, remove baseboards for a positive check for fire travel through walls and partitions. Any doubtful area should be wet down and checked again later.

Older walls, constructed of wood lath and plaster, are more susceptible to hidden fire than newer walls made with metal lath or wallboard. The old wood is extremely dry and easily ignited, owing to its rough finish. Older walls also have a wide inner space to accommodate wood joists. Walls of this type must be thoroughly inspected.

If a wall or ceiling contains insulation, both sides of the insulation must be checked. When the interior wall is opened, the paper or aluminum vapor barrier will be exposed. If no sign of fire is found, the outer side can be checked by removing some of the insulating material. The insulation can often be removed in sections and easily replaced with little damage.

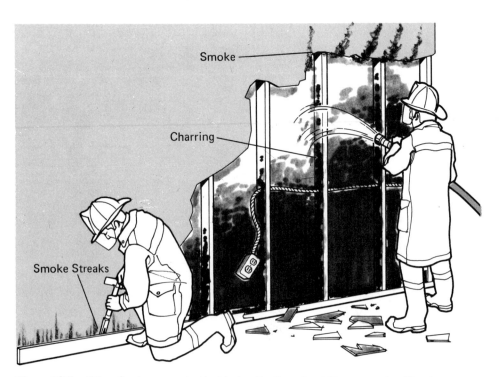

Figure 13.7. When fire is suspected behind walls, they should be opened until a clean area is found and the area should be wet down with a light stream if necessary.

When truck crews find that fire has penetrated a ceiling space, they must assume that it has spread into the floor above. That floor must be checked carefully. Full examination of the area might require the removal of flooring. This is especially important along walls and partitions and where shafts pass through the floor, usually in a corner of a room. Again, if part of the flooring must be removed, truck crews should take it up until a clean area shows the full extent of the fire. In order to hold damage to a minimum, the flooring should be removed as cleanly as possible. Cutting with power saws may be better than chopping with axes at this stage.

In general, cutting during overhaul should be done with power saws rather than with axes and other hand tools. The saws cause much less vibration than hand tools and thus allow fire fighters a greater margin of safety. This is especially important in a building that has suffered structural damage and requires extensive overhaul work.

Shafts. When truck crews suspect or find that fire has spread into a vertical shaft, the shaft must be opened and checked (Figure 13.8). Any natural openings into the shaft should be used for this purpose; otherwise, the shaft must be cut, then checked for signs of fire and smoke. When necessary, a stream or streams should be directed up into the shaft to wet down possible hot spots. In this case, the shaft opening might have to be enlarged to allow streams to be manipulated properly. In addition, truck crews should be detailed to check the top and bottom of the shaft for fire and sparks.

Shafts that were opened for venting or fire control during fire fighting operations must be thoroughly checked out at this time. Even though the fire has apparently been extinguished, the intense heat that was confined within the shaft might have rekindled a fire. For the same reason, anything in contact with these shafts (such as floors, walls, attic spaces or the roof) must be thoroughly inspected. Floors, walls and ceilings that abut the shaft might have to be opened to assure a thorough check. If there is any sign of heat or fire extension, the floor, wall or ceiling should be opened further, until a clean, untouched area is found.

Cabinets and compartments. The necessity of checking for fire extension around built-in cabinets and compartments located over or near fire areas was discussed in Chapter 6. If such cabinets have been subjected to fire or intense heat, they must be thoroughly checked during overhaul, whether or not they were opened previously. Otherwise, small fires might remain in the spaces between the cabinet shelves and the floor.

Window and door facings. When it is apparent that fire has involved window or door facings, truck crews should remove the facings and check the concealed recesses for fire (Figure 13.9). If the facings are in good condition, remove them without damage using an axe or pry tool, so they can be replaced when repairs are made.

If fire extension is found when the facings are removed, the walls or partitions must be opened to the end of the fire travel, and the area wet down as needed. Wainscoting is handled similarly; it is removed until a clean area is found, and the area is wet down if necessary.

Basement areas. When fire has directly involved a basement or cellar, that area must be checked completely, the same as any other area of the fire building. Even when a basement area has not been involved, check there for fire that might have fallen from upper levels. This is especially important

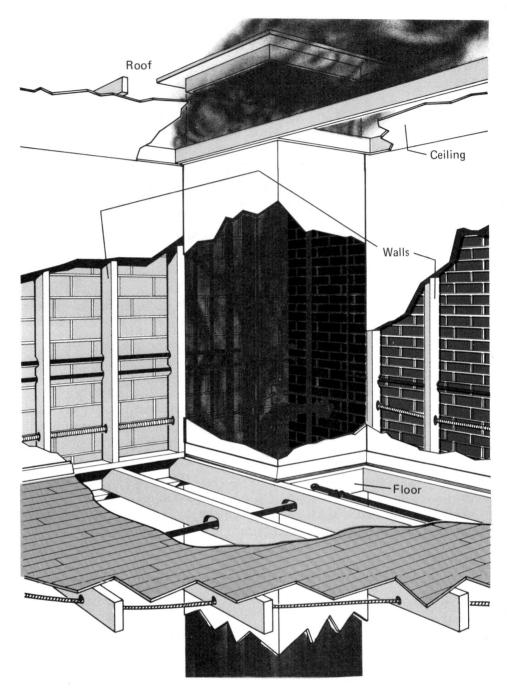

Figure 13.8. When a shaft has been the avenue of travel for fire, the areas adjacent to the shaft must be opened.

when it is known that fire traveled through vertical shafts that are open at the basement level.

Party walls between adjoining structures also must be examined carefully. Both sides of such walls must be checked, with special attention given to the points at which joists from the two adjoining buildings overlap or abut. Embers that easily lodge at these points and continue to smolder are difficult to find unless a careful check is made.

Chemicals and Other Hazards

Fires in chemical warehouses, drugstores, and cargo in transit can make overhaul operations extremely dangerous. If possible, fire fighters should know, prior to overhaul, which chemicals and flammable liquids are kept in storage areas. In some cases, prefire planning surveys are of assistance; more often, however, the type of material stored or in transit is constantly changing, especially in general warehouses and freight depots. This situation makes positive identification difficult or impossible.

The characteristics of combustible materials and chemicals vary in many ways. For example, some are heavier than air, and others are lighter than air; they may have high or low ignition temperatures. Chances are, however, that leaking flammables will be ignited if an ignition source is present. Therefore, care must be exercised during all fire fighting operations in this type of occupancy. Especially during overhaul, attention must be paid to labels on containers and to signs posted in and around the fire structure.

Protective clothing is essential for all fire fighters engaged in overhaul. Helmets, eyeshields, coats, gloves and boots should be worn for protection from dangerous chemicals. Self-contained breathing apparatus should be

Figure 13.9. When fire has been burning in a closed area, window and door facings must be carefully removed to check for hidden fire.

worn in hazardous atmospheres; nonpoisonous chemicals can leak, mix and form poisonous fumes in areas where fire fighters do not expect them.

Radioactive materials might be stored in some occupancies, their presence not known until the overhaul operation has begun. When such material is discovered, the area should be cleared, and the health department or other government agency responsible for dealing with such materials should be notified. Take readings to determine the intensity of radioactive emissions. Fire fighters must be carefully and thoroughly checked. If they have been exposed to excessive radiation, proper medical attention must follow, and their gear and tools must be decontaminated according to standard procedures.

Searching for the Cause of Fire

One important part of overhaul is the discovery and preservation of evidence of arson (Figure 13.10). Although not all fire fighters are arson investigators, they should be trained to look for signs of a deliberately set fire. When truck company personnel jump into overhaul without checking the fire building for such signs, the evidence might accidently be thrown out of the building, buried, or washed away.

One indication of a deliberately set fire is its location, especially if it started at the bottom of a stairwell or shaft. Other things to check for include multiple fires, unusual odors, undue wood charring, uneven burning, holes made in walls and floors, heating equipment not in proper condition, empty accelerant containers, residues of wax or paraffin, opened or removed service doors or panels to shafts, and inoperative sprinkler systems, fire doors, or other protective devices.

When a number of suspicious fires have occurred in an area, that information should be sent to the responding companies to alert fire fighters. They, in turn, should be especially alert for the deliberately set fire and should call investigators to the fire scene when anything unusual is discovered. This way, although a particular fire company may operate at only one of the suspicious fires, investigators who have been to several can look for similarities among the fires which might be used to establish a case of arson.

Restoration and Protection

When overhaul operations have been completed, the building and its contents should be restored and protected, as much as possible, from the elements and from vandalism.

The building. If vertical ventilation was required, skylights might have been removed, scuttles cut open, or holes cut in the roof. These must be closed as much as possible before the fire department leaves the scene.

A skylight that was removed without damaging the glass or frame can be replaced in its original position. Cover skylights with a salvage cover or plastic sheet if a good seal cannot be established at the roof line.

Large holes in a flat roof can be covered with salvage covers or plastic sheets. First, a ridge board (and other boards as necessary) should be placed to allow water to run off the roof and to keep the cover material from falling or sagging into the opening. The roofing material can be turned up, and propped up if necessary, to prevent water from running into the opening. A salvage cover should not be nailed to the roof. Instead, each end should be wrapped around a board that is longer than the cover and nails driven through the

Figure 13.10. Evidence of arson must be preserved.

protruding board ends into the roof. Both ends of the cover should be secured in this way for complete protection.

Roof drains must be cleared of debris and kept open; otherwise, in a heavy rain, the roof water level will rise, and water will seep into the roof holes.

Openings in pitched roofs can be covered with salvage covers, tarpaper or plastic, depending on their size. The roofing material above the hole should overlap the top of the cover, which should be held in place with wood strapping or slats.

Furnishings and stock. Items removed from the building during the course of fire fighting operations or during overhaul should be returned to the building if possible. If the items must remain outside, place salvage covers over them, tied down securely. The fire department should arrange for the security of these items (and their covers) and have the owners relieve the fire department of responsibility in writing.

When items that are returned to the building could be subjected to damage from dripping water or weather (owing to the condition of the building), these items also should be covered. If the building cannot be locked securely, make arrangements for protection or for a release from responsibility.

SUMMARY

The main purpose of overhaul is to ensure that the fire is completely out. For this, the fire structure must be thoroughly examined for smoke, embers, sparks, hot spots and fire. Truck crews must find and expose anything that could rekindle another fire, and engine crews must wet down any area that shows signs of fire or smoke. These operations should be performed as carefully and as safely as possible, with a minimum of additional damage to the building and its contents.

To reduce the chance of injuries:

- Conduct a preinspection before overhaul begins
- Mark and barricade hazardous areas
- Assign fresh personnel to overhaul duties
- See that overhaul personnel work as companies or in groups
- Observe special precautions in structures in which flammable or hazardous materials are stored.

GLOSSARY

Adz—cutting tool (chisel) with blade set at right angle to handle.

Aerial Apparatus—fire apparatus with a permanently mounted aerial ladder or elevating platform in addition to its regular equipment.

Aerial Company (*see* Truck Company)

Aerial Ladder—power-operated ladder of two or more sections that is permanently mounted on a fire apparatus.

Aerial Platform—mechanically or hydraulically raised platform mounted on a fire truck. Used for rescue and fire fighting, the platform can have an articulated (folding) boom, a telescopic boom, or some combination of the two. (*also called* Articulated Platform, Elevated Platform, Extending Platform, Snorkel)

Articulated Platform (*see* Aerial Platform)

Backdraft—explosion or rapid burning that can result from sudden introduction of oxygen into fire in a confined space.

Bangor Ladder—extension ladder equipped with staypoles that are used to assist in raising the ladder and to help provide stability; one staypole is attached to each beam of the ladder's base section.

Bars—prying tools used by a truck company member to force entrance or ventilate the fire floor. (*also called* Irons)

Base Section (*see* Bed Section)

Basket—riding/working platform on aerial apparatus. (*also see* Stokes Litter)

Battering Ram—large pole with larger head, generally made of metal, weighing about 50 to 60 lbs. and equipped with handholds; used to beat down doors, walls, etc. The other end can be pointed or have chisel points.

Beam—rigid or trussed structural side member of a ladder that supports the rungs or rung blocks.

Bed Section—largest section of an extension ladder; it acts as the support for the upper section(s) when the ladder is in the raised position. (*also called* Base Section)

Bolt Cutter—shears used for cutting bolts, links of chain, etc.; they can be powered by hand, air, or a hydraulic mechanism.

Brace (*see* Floor Lock)

Building Code—ordinance(s) regulating construction of buildings and such details as classification of structures, areas, heights, fire resistances, fire stopping.

Busbar—short conductor forming a common junction between two or more electrical circuits.

Butt—foot of lower end of a ladder.

Carbon Monoxide—colorless gas (CO), lighter than air and highly toxic by inhalation, with an ignition temperature of 1128°F and an explosive range of 12.5 to 74 percent.

Casement Window—one or two sashes hinged on the side so glass swings outward; a crank or other operating mechanism is located inside the window.

Catchall—basin made from a salvage cover to catch and hold water dripping from a ceiling.

Chock Blocks (*see* Chocks)

Chocks—blocks used to secure the wheels of a fire department or other vehicle to prevent movement. (*also called* Chock Blocks, Wheel Blocks)

Claw Tool—metal forcible-entry device with a hook and fulcrum at one end and a claw blade at the other.

Cockloft—small, usually vacant, loft or attic space between the top story ceiling and the roof of a building.

Combustion—rapid exothermic oxidation process accompanied by continuous evolution of heat and, usually, light.

Common Wall (*see* Party Wall)

Conduction—transmission of heat by means of a conductor or by direct contact with a heated element.

Confine—keep fire from spreading.

Convection—transfer of heat that occurs because of mixing or circulating of heated fluid.

Crotch Pole (*see* "U" Pole)

Damper—valve or plate for regulating air movement through a flue.

Double-Hung Window—window with an upper and lower sash or frame, each able to move vertically independent of the other.

Double-Pane Glass—two panes of glass with an air space between them for insulation.

Elbow—coupling with an angle, used to eliminate hose kinkings.

Elevated Platform (*see* Aerial Platform)

Elevated Stream—defensive fire stream provided by a ladder-pipe, or a master stream appliance on an elevating platform or boom.

Elevator Housing—small enclosed structure, usually on the roof, that protects elevator operating machinery at the top of an elevator shaft.

Ember—glowing fragment from a fire; it is capable of igniting other fires and is particularly dangerous when wind-blown.

Emergency Raise (*see* Hotel Raise)

Engine Company—basic unit of fire attack consisting of apparatus and personnel trained and equipped to provide water supply, water application from hose lines, location and removal of endangered occupants, and treatment of the injured.

Explosive Tool—shaped explosive charge which, when triggered, exerts all its force in one direction.

Exposure—property endangered by fire in another property.

Exposure Fire—fire which has spread from the exterior of one building to the exterior of another, or which has spread from one independent part of a building through a fire resistive barrier to another part.

Extending Platform (*see* Aerial Platform)

Extension Ladder—ground ladder of two or more sections that can be extended to various heights.

Face Shield—protective transparent device attached to the front of a helmet to protect the face from flying objects.

Facing—outer layer or coating applied to a surface to protect or decorate it.

Fire Building—structure in which there already is a fire, and from which fire might spread to other property.

Fire Tetrahedron—model of the four elements required by fire: fuel, heat, oxygen, and uninhibited chain reaction; each side is contiguous with the other three.

Fire Triangle—three-sided figure representing three of the four factors necessary for combustion: fuel, heat, and oxygen; it has now been replaced by the Fire Tetrahedron.

Fire Wall—wall of 3 test hours or longer fire resistance rating, built to permit complete burnout and collapse of the structure on one side without fire extension through or collapse of the wall itself.

Fireground—area of operations within which a fire is fought.

Flashing—sheets or strips of tin, copper or other metal placed in the roof and around chimneys and windows to weatherproof and prevent leakage.

Flashover—stage of a fire when all surfaces and objects are heated to their ignition temperatures, and flames break out simultaneously over the entire surface.

Flathead Axe—axe with a flat head that can be used as a sledge hammer or similar tool.

Floor Lock—device consisting of a heavy bar fastened to the floor inside a room and to a plate on the door; the bar slides into the plate for locking, and can be pivoted out of the way when the door is to be opened from the inside. (*also called* Brace, Police Lock)

Fly Section—upper section of an extension or aerial ladder.

Fog Stream—water stream in a fine spray pattern, produced by use of a fog nozzle.

Footplate—folding step at top of aerial ladder that provides a firmer footing than a rung.

Forcible Entry—activities required to remove barriers that keep fire fighters from a fire area.

Fox Lock—device with from two to eight bars that hold a door closed from the inside.

Free-Burning Fire—unconfined combustion.

Fuse Puller—insulated tool for handling fuses, especially for removing them from fuse boxes.

Garden Apartments—combustible low-rise multiple dwellings, including row houses and town houses.

Girder—large horizontal beam supporting walls, joists, or framing above the foundation.

Ground Ladder—ladder not mechanically or physically attached permanently to fire apparatus, and not requiring mechanical power from the apparatus for ladder use and operations. (*also called* Portable Ladder, Straight Ladder, Wall Ladder)

Halligan Tool—forcible-entry device which has a claw at one end and a spike and tapered pry head at right angles at the other end.

Halyard—rope or cable on an extension ladder used to raise the fly section(s).

Hammerhead Pick—driving device with a long, sharp-pointed pick used to dig dirt, concrete or debris.

Handline—2½-inch or smaller hand-held hose line that can be maneuvered while in operation.

Heavy Stream—any of a variety of heavy, large-caliber water streams, usually formed by siamesing two or more lines into an appliance with a large-diameter tip. (*also called* Master Stream)

Heavy-Stream Device—appliance designed to deliver a large volume of water through a nozzle with a large-diameter tip.

Hook and Ladder Company—name formerly used for a ladder or truck company.

Hose Roller—device with rollers designed to reduce the chafing of fire hose; when in place over a window sill, parapet, or ladder rung, hose can be dragged over the device without being damaged.

Hose Strap—strap with a handle used to secure or carry hose.

Hot Spot—any particularly hot location where it is difficult to extinquish a fire.

Hot Stick—insulated long pole with a hook used to move energized wire.

Hotel Raise—method of raising a ladder vertically so that there is simultaneous access to it from the interior of the building through windows on several floors. (*also called* Emergency Raise)

House Line—hose and nozzle attached to the outlet on a wet standpipe system.

Hux Bar—forcible-entry tool that resembles a small crow bar or claw tool; it is curved at one end and has a claw tool at the other.

Hydraulic Platform (*see* Aerial Platform)

Irons (*see* Bars)

Jack—portable device used to lift heavy objects by means of force applied with a lever, screw, or hydraulic press.

Joists—parallel horizontal beams laid wall-to-wall to support a floor or ceiling.

K Tool—forcible-entry device used to remove cylinder locks.

Kelly Tool—forcible-entry tool similar to a claw tool, with an axe head or blade at one end and a forked blade at the other.

Ladder Company (*see* Truck Company)

Ladder-Pipe—heavy stream appliance (nozzle) attached to the tip of an aerial ladder.

Laddering—positioning ladders for use in fire fighting and rescue work.

Lath—thin narrow strip of wood nailed to rafters or studs as a base for tile or plaster.

Lock Puller—device used to remove cylinder locks.

Masonry—term applied to anything constructed of stone, brick, tiles, cement, concrete, and similar materials.

Master Stream (*see* Heavy Stream)

Maul—heavy, long-handled hammer used to drive stakes, piles and wedges.

Nozzle—cylindrical constricting device, usually adjustable, attached to a hose to increase water velocity and form a stream.

Occupancy—use or intended use of an entire building or part of it.

Overhaul—job of searching for embers, sparks, small concealed flames, and any other remnants of a fire to make sure the fire is out completely.

Party Wall—wall common to two buildings. (*also called* Common Wall, Separating Wall)

Penthouse—enclosure on the roof of a building, ranging in size from a small stairway enclosure to a series of rooms large enough to use as living quarters.

Pickhead Axe—axe with a blade on one side of the head and a pointed pick on the other.

Pike Pole—wooden or fiberglass pole with a metal point and hook, usually used to pull sheet rock and plaster from ceilings.

Pitched Roof—roof elevated in the center or elsewhere to form a pitch (incline) to the edges; any roof with a more pronounced downward incline than a flat roof.

Plate Glass—strong rolled and polished glass with few impurities; used for large windows.

Platform Pipe—master-stream device mounted in the basket of an aerial platform and supplied with water by a system of pipes from ground level.

Plenum—space used in lieu of registers and ducts to collect used air and return it for recirculation.

Pole Ladder—folding ladder.

Police Lock (*see* Floor Lock)

Portable Ladder (*see* Ground Ladder)

Portable Standpipe—use of aerial apparatus to provide water for interior handline fire fighting operations: hose that has been carried or lifted into a fire building is attached to an aerial platform's 2½-inch outlet or master-stream nozzle connection; or connected to 2½- or 3-inch hose on an aerial ladder.

Porta-Power Unit—hydraulic jack device that can push and pull.

Prefire Planning—inspecting properties and developing written plans with drawings for fire fighting operations at specific properties or locations.

Preinspection—checking of a fire building by officers to make sure it is safe for overhaul before any fire fighters are sent into the structure.

Products of Combustion—gases, solid and liquid particulate matter, and residues that result from the burning process; visible products of combustion make up smoke, which usually has some degree of toxicity.

Pry Axe—multi-purpose prying tool with handle that can be extended or repositioned for leverage.

Pry Tool—any of a variety of prying devices used by the fire service to break locks, open doors, force windows, and pry up anything.

Pumper—fire apparatus that has a pumping capacity of 500 gpm or more, plus hose, ladders, and other equipment.

Rabbit Tool—small hydraulic spreader used to open doors.

Radiation—transfer of energy, including heat, through space by electromagnetic waves.

Rekindle—fire reignited after its apparent extinguishment.

Ridge Board—a horizontal timber at the upper end of the roof rafters, to which these rafters are nailed.

Roof Hatch—scuttle that extends from top story through the roof.

Rope Hose Tool—short length of rope with a ring at one end and a hook at the other, used for securing hose, etc.

Row Buildings—series of structures, usually identical, situated side by side and joined to each other by party walls.

SCBA (*see* Self-Contained Breathing Apparatus)

Salvage—procedures designed to reduce fire, smoke, water, exposure, and other damage before, during, and after a fire.

Salvage Cover—large sheet of waterproof material, available in many sizes, materials and shapes; sometimes fire resistant.

Scupper—drain placed at the base of a wall, even with the floor.

Scuttle—opening in the roof or ceiling of a building to provide access to the roof or attic.

Self-Contained Breathing Apparatus (SCBA)—equipment worn by fire fighters to provide respiratory protection in a hazardous environment; it consists of a facepiece, a regulating or control device, an air or oxygen supply, and a harness assembly.

Separating Wall (*see* Party Wall)

Siamese Connection—fitting used to combine two or more hose lines into one; the inlets are female and the outlet is male.

Single Ladder (*see* Straight Ladder)

Size-up—initial evaluation of a fire situation which forms the basis for fire attack.

Smoke Detector—device that actuates when it senses visible or invisible particles of combustion.

Smoke Ejector—fan used to remove smoke or hazardous gases from an area or to blow in fresh air to expel smoke, gases and heat.

Smolder—to burn and smoke without flame.

Snorkel (*see* Aerial Platform)

Solid Stream—hose stream that stays together as a solid cylindrical mass, as opposed to a fog or spray stream.

Spotting—placement of fire companies or fire fighting equipment at a fire for effective operation and attack.

Standpipe—wet or dry pipe, usually vertical, used to supply water to fire hose outlets in structures.

Stokes Basket (*see* Stokes Litter)

Stokes Litter—oblong basket stretcher made of wire or plastic, used to remove patients from heights or transport them over difficult terrain or debris. (*also called* Stokes Basket)

Straight Ladder—ladder with just one section. (*also called* Ground Ladder, Single Ladder, Wall Ladder)

Straight Stream (*see* Solid Stream)

Tempered Glass—glass that has been heat-treated to give it about five times the strength of ordinary glass.

Thermal Updraft—physical upward movement of air, caused by a fire, that produces a column of smoke and hot gases.

Thermostat—automatic control actuated by temperature change.

Tip—uppermost end of a ladder.

Truck Company—basic unit of fire attack consisting of apparatus and personnel trained and equipped to locate, protect and remove fire victims; provide forcible entry and gain access; ventilate; check for fire extension; control utilities; operate elevated master streams; and conduct salvage and overhaul operations. (*also called* Ladder Company)

Truss—1. combination of triangles that form a frame for supporting loads over a span. 2. side member of a trussed ladder.

"U" Pole—pole with a U-shaped fixture at one end, used to help erect and brace ladders. (*also called* Crotch Pole)

Ventilation—1. circulation of air in any space by natural convection, or by fans blowing air into or exhausting air out of a structure. 2. opening of doors and windows or holes in the roof of a burning building to allow smoke and heat to escape.

Venturi Siphon—dewatering device which uses a flow of water through it to propel liquids through a hose.

Wainscoting—material used to cover lower part of interior walls.

Wall Ladder (*see* Ground Ladder)

Water Thief—appliance that permits use of one 2½-inch line and two 1½- or 1¾-inch lines from a single 2½- or 3-inch supply line.

Wheel Blocks (*see* Chocks)

Window Lights—panes of glass set above or around a much larger plate of glass such as a store display window.

Working Fire—fire that requires fire fighting activity by most or all of the fire department personnel assigned to respond to the initial alarm.

INDEX

A

Aerial apparatus
  evolution of 3
Aerial ladder
  placing of 119
Aerial operations 113
  hose 128
  removing people 121
  rescue 115
  safe procedure for 114
  spotting the turntable 116
  ventilation 125
Aerial pipes
  rigging of 165
  setting up 161
Aerial platform
  placing of 120
Aerial streams
  choice of 173
  directing 173
  for exposure protection 171, 173
  for fire attack 167
  improper use of 170
  nozzles 167
  placement of 167
  updraft effects 168
  on weakened structures 189
Air circulating systems 180
Air conditioning
  problems with 177
Air shafts
  in fire venting 53
Apartments
  forcible entry to 106
Apparatus
  positioning of 16
Attic fires
  venting of 61
Axe
  flathead 10
  pickhead 10

B

Backdraft
  described 6, 76
"Bar" person 11
Basement
  as rekindling area 197
Basement entrance
  sidewalk 109

Basement fires
  in large structures 68
  venting of 61, 68
Blower
  forced-air 178
Building contents
  protection of 150
Building, multiple-use
  fire venting in 61

C

Cabinets
  as rekindling areas 197
Carbon monoxide
  dangers of 7
Catchalls 152
Ceilings
  as rekindling areas 195
Chemicals
  as overhaul hazards 198
Chutes
  for water removal 154
Claw tool 10
Combination occupancies
  forcible entry to 109
Commercial occupancies
  forcible entry to 104
Conduction
  dangers of 6
  defined 5
Conduits
  for water removal 151
Connection
  defined 2
Convection cycle 3
Cooling systems 180
Covers
  salvage 149

D

Dampers
  forced-air systems 179
Doors
  checking in search 34
  forcible entry through 102
Drains
  for water removal 156
Dwellings
  forcible entry to 106
  single-family
    rescue in 23

E

Electric service 184
  commercial structures 185
  dwellings 185
  industrial structures 185
  main switches 185
Electric wires
  problems with 177
Elevated handlines
  from an aerial ladder 175
  from a platform 174
Elevated streams
  for buildings 162
  for flammable liquid
    facilities 164
  for open storage areas 162
  value of 161
  water supply 164
Elevators 186
  for fire floor approach 72
  in fire venting 52
Entry, forcible
  (see Forcible entry)
Exposure hazards 172
Exposure protection
  aerial streams for 171

F

Facings
  as rekindling areas 197
Factories
  forcible entry to 109
Fans
  placement of 57
Fire
  smoldering 5
    indications of 74
    venting of 76
Fire cause
  searching for 200
Fire extension
  attached structures 89
  checking 79
    as truck company duty 7
  in concealed spaces 80
  exterior exposures 90
  horizontal 85
  interior
    open 89
    vertical 81
  walls

209

Fire resistant structures
   venting of 71
Fire spread
   mechanics of 2
Fireground
   approach to 16
   coverage of 12
      apparatus positioning 16
      front 12
      interior 15
      other aspects of 14
      rear 12
      size-up information 14
      truck out of quarters 14
Flashover
   defined 6
Flathead axe 10
Fog streams
   for fire venting 57
Forced-air systems
   problems with 178
Forcible entry
   apartments 106
   combination occupancies 109
   commercial occupancies 104
   through doors 102
   dwellings 106
   exposed buildings 96
   factories 109
   the fire building 95
   office buildings 109
   prefire inspection for 93
   tools for 10
      cutting 98
      explosive 99
      forcing 98
      lock pullers 99
      prying 98
      use of 98
   warehouses 109
   through windows 99

G

Gas
   bottled 183
   city 182
   meters 183
   units 182
Ground ladders
   for advancing hoselines 141
   bridging techniques 138
   climbing of 143
   emergency raise 137

   as exits 137
   handling 132
   hotel raise 137
   normal raise 135
   operations 131
   for positioning hose streams 143
   from roofs 142
   safety 135
   uses 131, 144
   in ventilation 140

H

Halligan tool 10
Handlines
   elevated
      (*see* Elevated handlines)
Hand tools
   flathead axe 10
   K tool 10
   lock pullers 10
   pickhead axe 10
   pike pole 10
Hatches
   roof 47
Heat
   forced-air 179
   kerosene 182
   radiant 4
   transmission of 2
Heating units 181
Hose operations
   from aerial apparatus 128
Hospitals
   rescue in 25
Hot stick 185
Hotels
   rescue in 24

I

Industrial occupancies
   rescue in 25
Inspection
   prefire
      for fire venting 54
      for forcible entry 93
Institutions
   rescue in 25
"Irons" person 11

K

K tool 10
Kelly tool 10
Kerosene heaters 182

L

Ladder-pipe operation
   as truck company duty 7
Ladder truck
   positioning of 17
Ladder work 132
Laddering
   as truck company duty 7
Ladders
   ground, size and type 9
Large structures
   basement fires in 68
Lifeline anchor
   aerial apparatus as 124
Litter
   for rescue 122
Lock pullers 10

M

Machinery covers
   in fire venting 52
Motels
   rescue in 24

N

Nozzles
   for aerial streams 167

O

Office buildings
   forcible entry to 109
Oil burners
   described 181
   problems with 177
One-story buildings
   ventilation operations 66
Overhaul
   building protection 200
   chemical hazards 198
   contents protection 201
   described 189
   fire cause 200
   personnel 190
   personnel control 191
   preinspection 190
   procedure 192
   rekindling 194
   as truck company duty 7
   work assignments 192

P

Penthouses
   in fire venting 51

Personnel
  positioning of 12
  for truck company 9
Pickhead axe 10
Pike pole 10
Porta-power unit 11
Power tools 10

Q

Quic-Bar tool 10

R

Rabbit tool 11
Radiation
  defined 4
Rekindling
  areas of 195
  indications of 194
Rescue
  aerial 115
  in apartment houses 23
  chronology of 20
  complexity of 19
  considerations in 22
  with ground ladders 135
  in hospitals 25
  in hotels and motels 24
  in industrial occupancies 25
  in institutions 25
  in residential occupancies 23
  in retail stores 27
  in schools 25
  in single-family dwellings 23
  as truck company duty 7
Rescue operations
  at alarm 20
  before alarm 20
  on the fireground 21
  immediate 21
  placing streams 22
  search 22
  size-up 21
  ventilation 22
Residential occupancies
  rescue in 23
Response procedure
  need for 12
Retail stores
  rescue in 27
Roof operations
  access 62
  cutting through 55
  personnel 63
  venting 64

Rope hose tool 129
Row stores
  ventilation operations 66

S

Salvage
  objectives of 148
  as truck company duty 27
  types of 147
Salvage covers 149
Schools
  rescue in 25
Scuttles
  roof 47
Search
  areas to be searched 32
  completion 32
  duties of truck crews in 29
  importance of 28
  pattern for 29
  standard procedure 29
  techniques for 34
  victims
    checking for 35
  visibility in 36
Sewer pipes for water removal 156
Shafts
  as rekindling areas 197
Shopping centers
  ventilation operations 66
Skylights
  in fire venting 44
Smoke ejectors 56
  placement of 57
Standpipe
  portable 130
Stokes basket
  for rescue 122
Stretcher
  for rescue 122

T

Toilets
  for water removal 156
Tools
  assignment of 11
  forcible entry 10
  hand 10
    claw 10
    halligan 10
    kelly 10
    Quic-Bar 10
  porta-power unit 11
  power 10

rabbit 11
training for use of 10
truck company 9
Training
  for tool use 10
Truck company
  aerial company
    as synonym 7
  duties of 7
  hook and ladder company as synonym 7
  initial assignments 9
  ladder company as synonym 7
  operations 7
  personnel 9
  role of on fireground 1
  snorkel company as synonym 7
  tools 9
Truck work
  objectives of 2
Turntable
  aerial, spotting of 116

U

Utilities
  control of 177
    prefire planning for 177
  as truck company duty 7
  operating controls for 178
  problems of 177

V

Ventilation
  basic principles of 40
  convection as a reason for 3
  effects of wind 42
  fog streams in 57
  forced 56
  natural openings 41
  prefire inspection 54
  in rescue operations 22
  roofs
    cutting through 55
  roof openings 43
  techniques of 39
  as truck company duty 7
  use of aerial equipment for 125
  value of 39
Ventilation operations 59
  adjoining buildings 66
  attic fires 61
  basement fires 61
  fire resistant structures 71

for venting 48
ground floor stores 65
for large structures 68
multiple-use buildings 61
one-story buildings 66
one-story dwellings 59
removing of 49
roof 62
row stores 66
shopping centers 66
two-story dwellings 60
Victims
  in search of 35

search conditions 36

W

Wall openings
  for water removal 157
Walls
  as rekindling areas 195
Warehouses
  forcible entry into 109
Water
  removal from buildings 154
  problems with 177
Water flow

controlling
  in salvage 151
Water pipes 186
  boilers and heating units 186
  shut-off valves 186
Water removal
  (*see* Salvage operations)
Wind
  effects of
    on ventilation 42
Windows
  in fire venting 41
  forcible entry through 99